Vortex Publishing LLC.
4101 Tates Creek Centre Dr
Suite 150- PMB 286
Lexington, KY 40517

www.vortextheory.com

© Copyright 2019 Vortex Publishing

All rights reserved. No part of this book may be reproduced or transmitted in any form or by any means, electronic or mechanical, including photocopying, recording or by any information storage and retrievable system without the prior written permission by the Publisher. For permission requests, contact the publisher.

Printed in the United States of America

1 2 3 4 5 6 7 8 9 10

Library of Congress Control Number: 2019953549

ISBN   978-1-7341843-0-3
eISBN 978-1-7341843-1-0

Editor's note: *All drawings in this book are original illustrations made by Dr. Moon. They are kept as they are to maintain the integrity of his work.*

TABLE OF CONTENTS

Forward From The Publisher..III
Welcome again……...IV
Introduction..V

PART I
THE HOLY GRAIL OF SCIENCE

Chapter 1: Does a Complete Theory of Everything Really Exist?...1
Chapter 2: The Entire Universe Is One Gigantic Particle..2
Chapter 3: The Discovery of Quantum Gravity Unites the Two Conflicting Theories................4
Chapter 4: It Is Space That Is Made of Something and Matter That Is Made of Nothing!.........7
Chapter 5: Einstein's Mistake Spawned Another Equally Egregious Mistake: "Particle Logic"....9

PART II
SPACE CREATES EVERYTHING IN THE UNIVERSE!

Chapter 6: What Is Space Made of?..10
Chapter 7: Time Does Not Exist; Hence, Space Does Not Possess Time Characteristics!!!.........13
Chapter 8: It All Begins With Three Dimensional Holes in Space!..17
Chapter 9: Space Is Like a Fabric That Can Rip Open Creating 3D Holes in Its Surface............21

PART III
THE PARTICLES OF NATURE

Chapter 10: The Secret of the Neutron..27
Chapter 11: The Two Sides of Space...32
Chapter 12: Quarks Are Higher Dimensional Holes in Space..35
Chapter 13: The "Particles" of Nature Containing Quarks Are Briefly Explained.....................39

PART IV
THE "FIVE" FORCES OF NATURE!

Chapter 14: The Strong Force...42
Chapter 15: The Weak Force...45
Chapter 16: The Electromagnetic Force..48
Chapter 17: The Force of Gravity IS A PUSH NOT A PULL!!!..53
Chapter 18: The Discovery of the Fifth Force in Nature: The Anti-gravity Force!!!..................57

PART V
THE EXPLANATION OF ENERGY

Chapter 19: The True Explanation of the Photon..60
Chapter 20: Maxwell's Mistake: He Described the Effect and NOT the Cause!.......................64
Chapter 21: The Alpha Particle..72
Chapter 22: The Pauli Exclusion Principle..73
Chapter 23: Gamma Rays and X-Rays..77

PART VI
STRANGE RELATIONSHIPS PREVIOUSLY EXPLAINED ONLY BY MATH ARE ILLUSTRATED!

Chapter 24: Planck's Constant 83
Chapter 25: Illustrating Einstein's Famous Equation [$E = mc^2$] 85
Chapter 26: Mass 86
Chapter 27: Newton's Law of Gravity: $F = G\, m_1\, m_2 / r^2$: And His 3 Laws of Motion
 & $F = ma$ 90
Chapter 28: Schrödinger's Equation: And the Foundation for All of Quantum Mechanics 96

PART VII
BIZARRE PHENOMENON EXPLAINED; AND THE EXPOSURE OF FALSE IDEAS

Chapter 29: The Explanation of the Double Slit Interference Patterns
 Created by Single Electrons and Photons! 102
Chapter 30: Dark Matter 106
Chapter 31: Dark Energy 108
Chapter 32: Boson Particles Do **NOT** Transfer the Forces of Nature!!! 113
Chapter 33: The Explanation of Quantum Entanglement 116
Chapter 34: How the Four Quantum Numbers of Atoms Are Created 126
Chapter 35: Radioactive Decay 131
Chapter 36: Beta Minus & Beta Plus Decay 133
Chapter 37: The Unification of Quantum Mechanics and Newtonian Physics via Nuclear
 Gravity! 140
Chapter 38: Summing It All Up! 143

PART VIII
THE APPENDIX

#1: Who says scientists do not believe in God??? 152
#2: List of revolutionary discoveries 158

REFERENCES

National/International Conferences attended, and peer reviewed scientific papers presented 162
Books by author {A} and work presented in other published books/booklets 166
Other references 167
Russian Scientific Journals 167

ALSO…another BLOCKBUSTER DISCOVERY…finally explained is the greatest mystery in all of Physical Science: the unification of Newtonian physics and Quantum Mechanics: something never before done in the annals of science!!!

Forward from the Publisher, Stan Clifford, CEO Vortex Publishing

Hello, I am Stan Clifford. I met Dr. Russell Moon eight years ago very randomly. A new friend of mine, Jaime Lemos, met a protege of Russell's named Fabian Calvo. Fabian had accompanied Russell on a trip to Russia where Russell created an experiment to increase the speed of light by 204 miles per hour. Fabian told Jaime about Russell and Jaime told me I should meet this physicist. I did and we decided to form a company, V.T.I., LLC with the mission to expose Russell's discoveries to the world at large and work on applying his knowledge in practical applications.

At that time, I understood what the concept of a theory of everything was. I knew it had been a lifelong pursuit of Albert Einstein and Stephen Hawking. I also knew they were both very far away from getting there. I also had no clue at that time that Russell's Vortex Theory could help enable the accomplishment of this massive feat.

Now I find myself as the CEO of the company that is printing the books, The Thesis of The Complete Theory of Everything.

Over the last eight years, I have been learning with amazement about Russell's science. When I tell my friends our company has discovered some of the most amazing things ever discovered by a scientist anywhere I get these blank stares looking at me. Really?? And I say, yes, really. Wait until all of the books come out!!

Dr. Moon has not only become a great business partner, but he has also become a great friend. I am in awe of his intellect and what he has done, and I am so proud to be helping him bring his science to the world.

Stan Clifford

2021

Welcome Again...

To all you good friends and to all you good friends I will never meet, welcome again: this time to the sixth book in what was originally meant to be only a three volume trilogy called *"The End of the Concept of Time I-III"*! However, brand new revolutionary discoveries made since the first three books required the publishing of other more up-to-date books, and this one is no exception: in fact it presents perhaps the last and greatest scientific discovery ever made in the history of mankind...

"The Complete Theory of Everything"!

As mentioned before, during the last thirty years of using the *Vortex Theory of Atomic Particles* to discover the explanations for over 120 great scientific mysteries of the universe, I began to notice a curious underlying pattern that slowly led to the shocking revelation that EVERYTHING in the physical universe appears to be created out of just one thing, one object!

I was stunned! Excited! Could it be true? I knew I had to prove it if it was, but the proof seemed impossible. It necessitated explaining four more seemingly impossible mysteries of physical science: quantum gravity; the unification of Quantum mechanics and Newtonian physics; the real cause for the creation of the electron sub-shells surrounding atoms; and the discovery of a 5th force in nature: the anti-gravity force?

But sometimes the impossible happens! The above discoveries were made, [and are in this book], and with them, the Theory of Everything was finally discovered! And everything means "Everything"; "Everything" that exists in the physical universe: matter space, time, energy, and the forces of nature are suddenly explained by only one thing, one object!

This book presents the evidence of this fantastic discovery by explaining such daunting, never explained mysteries as: *the Force of Gravity, Quantum Entanglement, the Constant of Fine Structure (the mystifying, dimensionless number of 1/137), Dark Energy, Dark Matter, the Asymmetric Parity of Neutrinos, the Pauli Exclusion Principle, how single photons create Double Slit Interference Patterns, the fifth force in nature: the Anti-gravity force; the unification of Quantum Mechanics and Newtonian Physics and how the Orbitals of electrons surrounding the Nuclei of atoms are created!!!*

Even more important, because of this great discovery, the mysteries of the universe that previously could only be explained by complex formulas can now be illustrated! And just like the children of today looking at textbook drawings and videos of the orbits of the planets know more about the construction of the Solar System than the greatest philosophers of the Golden Age of Greece; because of these drawings, for the first time ever, you and the children of all future generations will know more about the construction of the entire universe than ALL of the greatest scientists who have ever lived!!!

INTRODUCTION

While writing the explanation for one of the greatest mysteries in all of physical science: the Unification of Newtonian Physics (explaining the motions of the planets and stars); and Quantum Physics (explaining the mysterious workings of the invisible world of subatomic particles); my friend and colleague Mr. Stan Clifford, CEO, Deco Art, called me on the phone. He asked, "Is it possible using this great discovery to now also use *The Vortex Theory of Atomic Particles* to explain to High School students the knowledge of the physical universe that usually takes many years of study in university level scientific courses? Knowledge usually requiring an advanced level of both physics and mathematics to comprehend?"

"YES!"

He then asked, "How is that possible?"

I explained…[the answer went something like this]…"for thousands of years, the motions of the sun, moon, and planets were once great mysteries, whose explanations were sought after by the greatest minds of ancient history's most sophisticated, and educated individuals. Mysteries whose explanations were deemed too complex for the 'common man' to try to explain or comprehend!

"But today, all that has changed! Using simple drawings of planetary orbits in books, today's grammar school children can be taught and know more about these mysteries than any of the greatest philosophers of the Golden Age of Greece: Socrates, Aristotle, Pythagoras, Archimedes, and Plato!

"And a similar revolution is about to happen all over again with the discovery of *'The Complete Theory of Everything'*. Simple drawings can now explain scientific mysteries such as dark energy and dark matter that until now had no explanation whatsoever!

"Furthermore, because of the discovery of this Theory of Everything, simple drawings can now be substituted for mathematical formulas that used to be needed to explain complex scientific concepts. This allows the readers of this book, and the children of future generations, to know what previously was understood only by spending years in university mastering the disciplines of physics and mathematics.

MISUNDERSTOOD PHENOMENON CREAT MISAUNDERSTOOD EXPLANATIONS…

"Today's children know more because we now know that these so-called 'Great Mysteries of Astronomy' were actually phenomenon created by causes the people of the ancient past knew nothing about: causes such as the rotation of the Earth creating the phenomenon of night and day; and making it seem that everything in the sky appears to be orbiting around us, [and compounded by the Moon that actually does orbit the Earth!]…

…furthermore, the orbit of the Earth about the sun also creates the appearance of different star formations [constellations] emerging in the night sky at different times of the year: creating the 12 signs of the Zodiac. And, since the 23.5 degree tilt of the Earth was also unknown, ancient man did not know what was creating the regular appearance and disappearance of the four seasons.

"More sophisticated phenomenon, such as the elliptical orbits of the planets and their different velocities as they rotate around the sun, create the strange occurrence known as 'retrograde motion', and caused Ptolemy, the ancient Egyptian Astronomer of the Second Century AD, to conclude that the planets were sitting upon giant rotating wheels revolving about the Earth, a simply crazy idea that incredibly, lasted for 1300 years!

"The above revelation is what we used to call in the 60's, "A mind blower"! For how could generations of grown adults believe for 1300 years in the existence of gigantic wheels up in the sky they could not see? Were they just dumb, naive, or just unwilling to go against the so-called "educated" individuals of their eras expounding these crazy ideas?

"But then…are we any better? How come today's scientists who believe in Einstein's explanation for gravity being created by 'bent space', never question the ridiculous premise this incredibly foolish idea is based upon: space is made of nothing!!! At first it does not seem to be very important. But then, a little reflection causes us to take a closer look at Einstein's curious premise and suddenly we are left with an uneasy conclusion: *for how can something made of nothing be bent*? Remember, it was Einstein who said, "*Space is made of nothing*!" So again…

"*How can 'nothing' be bent*?"

"Although there are many other stellar phenomena such as eclipses of the sun and moon; and the discovery of the vernal and autumnal equinoxes; the above examples are enough to make the point that an individual does not have to be an astronomer possessing advanced mathematical knowledge to understand what is happening in the sky! Simple drawings and videos of the motions of the earth and planets in grammar school classrooms now explain it all!

STRANGE PHENOMENON CAN SPAWN SOME CRAZY EXPANATIONS…

"The exact same situation is about to happen with the workings of the sub-atomic universe. The explanations for the seemingly strange creations of colliding particles; and the seemingly impossible abilities of moving particles to pass through solid objects in the sub-atomic world are now easily explained, in fact, almost trivial!

[Note: today's scientific explanation for this incredible phenomenon is that the electron has the ability to change from a particle into a wave; pass through the solid wall as a wave then change back from a wave into a particle again then proceed along its way; WOW! This ridiculous notion almost seems to indicate that the little electron possesses consciousness! Or is it like the "Changeling" of mythology: a creature able to change its shape at will?

"Fortunately, with the discovery of the *Vortex Theory of Atomic Particles*, this explanation is revealed to be an outrageous, egregious mistake! Pure Foolishness! Later on, you will see the true explanation for this extraordinary phenomenon that has come to be known as "Tunneling".]

"Explaining the many discoveries of the *Vortex Theory of Atomic Particles* by using drawings, suddenly allows everyone to understand difficult subjects previously explained only by mathematics, and what was once exclusively reserved only for those possessing advanced degrees in Quantum Mechanics, Particle physics, or 'Quantum Chromodynamics' (what a word!). Shockingly, like a magic cipher, this Vortex Theory easily reveals knowledge once desperately sought after by men like: Einstein, Pauli, Heisenberg, Hawking, Eddington, Pauling, Jung, Sommerfeld, Debye, Bethe, Rabi, Max Von Laue, Bohr, Schrödinger, Max Planck, Richard Feynman, and many, many more!

"These great scientists and thinkers were limited in their abilities by the fact that they, like those famous philosophers of ancient history, were using incorrectly explained phenomenon to try to understand the causes for the seemingly bizarre happenings in the sub-atomic world!"

The revelation of this misunderstood scientific knowledge and phenomenon are important not only to the scientists of this era, but also, to each and every one of us too: because unknowingly, we use it to formulate our thinking processes and explain many other things that seem to be totally incompatible with science; such as our religious beliefs → as we shall all soon learn.

So whoever you are, and whatever you know now…you too have the right to know the truth…to take a look into the future…and know what one day will be common knowledge to the children and scientists of all future generations!

In honor of two of Russia's greatest scientists: Dr., Prof. Konstantin Gridnev; and Dr., Prof. Victor V. Vasiliev, I have given the vortices the name "Konsiliev Vortices".

What you are about to witness is for everyone, no physics or mathematical skills are needed!

Russell Moon

PART I
THE HOLY GRAIL OF SCIENCE!

> Discovering the Complete Theory of Everything is The Holy Grail of Science → and the Holy Grail of Science is the Complete Theory of Everything! Nothing, no other theory in the history of the world is more important!

Chapter 1
Does a Complete Theory of Everything Really Exist?

"Is the title of this Book true? Does a Complete Theory of Everything really exist?"

"Although it is hard to believe, the answer is yes!"

"The *Vortex Theory of Atomic Particles* has made many revolutionary discoveries, but perhaps the greatest of them all is the discovery that Everything…"Everything" that exists in the physical universe is created out of only one thing: one object!

In its ability to explain all of the great mysteries of the physical universe, the Vortex Theory also unexpectedly discovered the last and greatest mystery of them all: the unification of everything that exists in the universe! The thesis presented in this Book reveals the evidence of this great and fantastic discovery…

In the past, notable scientists like Albert Einstein & Stephen Hawking have sought to discover what they called the 'Theory of Everything': a theory referring to their attempt to unify the forces of nature. But this was a misleading, exaggerated description of what they were trying to discover. The universe consists of five parts: matter, space, time, energy, *and* the forces of nature. Hence, for one to try to unify the forces of nature so they can state they have discovered the Theory of Everything is a gross exaggeration that totally ignores the other four parts! A Theory of Everything is not selective. It means "Everything": "Everything" that exists in the physical universe; and that is exactly what this *Vortex Theory of Atomic Particles* does. It explains Everything, Everything that exists in the physical universe! It is what can only be called: *"The Complete Theory of Everything* (Ch. 38)!"

This theory now explains ALL of the great mysteries of physical science and the previous complex laws of nature that once could only be described using difficult to understand abstract concepts and higher mathematics. Even more amazing, all of these previously unexplainable mysteries can now be explained by using simple illustrations! Because of this fact, all readers of this thesis do not need to have any prior knowledge of physics or mathematics.

Consequently, the purpose of this book is to present this great knowledge to all people everywhere in the world, not just the scientific community. This book presents a visual journey through the physical universe using drawings to explain what once was an exclusive field, restricted only to those possessing extensive knowledge of physics and mathematics. As said before, no knowledge of physics or mathematics is needed!

When finished, the reader will know more about the physical universe than all of the greatest scientists who have ever lived!

> The Holy Grail of science begins with a revolutionary vision of the universe unlike anything anyone has ever imagined before: the entire universe is one gigantic particle! Even more fantastic, it is in the process of turning inside out: creating two volumes of space; one in expansion, one in contraction!

Chapter 2
The Entire Universe Is One Gigantic Particle!

The "Overview" from Book 4 revealed a shocking revolutionary vision of the universe unlike anything anyone has ever imagined before: the entire universe is revealed to be one gigantic particle! Even more fantastic, this particle is in the process of turning inside out: creating two volumes of space; one in expansion, one in contraction! (Book 3: Ch.5) (This effect can be envisioned by taking a large rubber balloon and slowly turning it inside out.)

This expansion and contraction can be more easily understood by using the analogy of two spheres – one inside of the other – and sharing the same volume. As the volume of the exterior sphere [red] flows into the interior sphere [green], the volume of the interior sphere increases and the volume of the exterior sphere decreases. As one cubic meter of space is added to the interior of the sphere, one cubic meter is subtracted from the volume of the exterior sphere. Note in the figures below, that as the interior volume increases, the exterior volume decreases.[1]

Figure 2.1

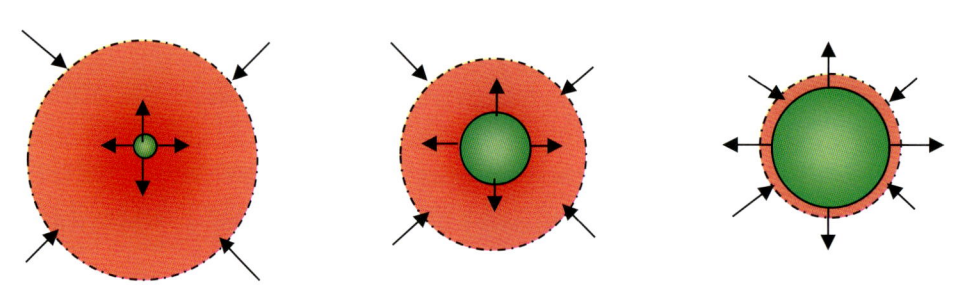

When originally discovered, it was not then realized what this vision of the universe really represented! Since then, it has been recognized for what it really is: "The Holy Grail" of science; "The Complete Theory of Everything"! A theory that explains Everything…"Everything that exists in the universe"!

The 20th Century science's explanation for the creation of the universe via the "Singularity" is now revealed to be a mistake. Presented here in this Book is the shocking proof for the real creation of matter and everything else in the universe being the creation of one gigantic multi-dimensional "particle!" What we call "space" is not the vast void we have all been mistakenly taught to believe; but instead, is actually the invisible substance this colossal particle is made of! This is not supposition but is based upon the thesis for which a PhD in nuclear physics was awarded in 2005,

---

[1] For scientists: This is also not speculation but is based upon the Asymmetric Parity of neutrinos and the ±charges created by decaying Quarks. See Book 3.

published by the Russian Academy of Sciences in 2012, and after three experiments revealed the physical evidence of this revolutionary thesis.[2]

The subatomic particles everything in the universe is made of are not little solid particles as is presently believed, but instead, are really microscopic three dimensional holes in space previously explained in Books 1-5, and then later again in this book (Ch. 8-9). An easy way to envision these tiny holes is to take a bottle of liquid soap, shake it, and then look at the bubbles floating in the fluid.

Although from our point of view, all matter appears to be separate from all other matter, this is only an illusion. Every infinitesimally small subatomic "particle" in the universe that atoms are constructed out of, including those in our bodies are really just part of the one massive, gigantic particle. We are all part of it; nothing is separate from anything else!

All is One!

But will it last forever?

In a way it will!…And in a way it won't!

Just as the beginning of the inside-out twist in space, [there was no "singularity" as 20th Century science proposed], started an instant expansion of one volume out the other in a single instant of time; creating what science mistakenly calls the "Big Bang"; the end will also occur in a single instant of time: "in the blink of an eye!"

When all of the space from the contracting volume has flowed into the expanding volume, the cycle will be complete; there will be no more space to feed the expanding volume, and all of the holes in space the matter of the universe is created out of will instantly disappear! Everything will be gone! Including us!

[So…"in a way it won't last forever."]

However, just after this instantaneous universal destruction ends, it will instantly begin all over again! The analysis of the distribution and different formations of the giant galaxy clusters in the universe reveals that this has happened many times before in the past (Book 4). The instant the contracting volume disappears, the cycle begins anew. A rip in space will instantaneously re-occur in the former expanding volume's geometric center, and another "Big Bang" will create another new universe. The space flowing into the rip will become the new expanding volume, and the former expanding volume will become the contracting volume!

[So…"in a way it will last forever!"]

Throughout the rest of this book, we will present the evidence of this shocking thesis!

---

[2] Dr. Russell Moon PhD thesis; The End of "Time", Book of Academic Papers, International Conference; Natural and Anthropogenic Aerosols VIII, October 1-5 2012, The Learned Works Addendum, Part 2, 2013 Saint-Petersburg Russia, 2013; pp 473-488; VVM Publishing House 2019.
ISBN 978-5-9651-0804-6

> The discovery of quantum gravity allows the first ever unification of two theories of the construction of the universe that were once believed to be absolutely diametrically opposed: Quantum Mechanics [the theory explaining the motions of sub-atomic particles]; and Newtonian Physics [the theory explaining the motions of the planets, stars, and galaxies].

Chapter 3
The Discovery of Quantum Gravity Unites the Two Conflicting Theories

As explained in Book 4, when the discoveries about what space was capable of creating began to become more and more apparent, I did not realize then what I had discovered. However, after more and more subsequent scientific discoveries were made using this Vortex Theory and its continuing revelations about what space was capable of creating, I began to realize that this one thing, this one object might be responsible for creating everything that exists in the entire physical universe! If true, it could in fact be the object responsible for creating the fabled *Theory of Everything*: the legendary theory sought after by all of the greatest philosophers and scientists who have ever lived!

But was it true?

Years ago, no matter how much I wanted to, this was not a question I could answer because there were still several other, major mysteries of the universe I needed to explain: such as the greatest of them all → the seemingly impossible unification of the two dominant but diametrically opposed theories of science: Quantum Mechanics and Newtonian Physics.

But I did not have to wait long. Because with the blockbuster discovery of quantum gravity, I suddenly knew I had also discovered the invisible link between Quantum Mechanics and Newtonian Physics (Ch. 37). Using this link, I discovered how to unify these two seemingly incompatible theories of the universe: Quantum Mechanics [the theory explaining the motions of sub-atomic particles]; and Newtonian Physics [the theory explaining the motions of the planets, stars, and galaxies].

When I discovered how to unify these two seemingly incongruent theories, the other last great secret of the universe was suddenly revealed (Ch. 38)…"The Theory of Everything", the theory I like to call: *The Complete Theory of Everything…The Holy Grail of Science!*

Like a magician's trick, this vision just instantly appeared, seemingly out of nowhere with the unification of these two theories. Because of this final discovery, my life, your life, and all of the lives of all the peoples of all future generations will never be the same again!"

This will happen, because with the discovery of how to unify these two seemingly contrary theories of the universe, Quantum Mechanics & Newtonian Physics comes the even greater, monumental discovery: that everything that exists in the physical universe is created out of just one 'Thing', one 'Object'!

THE COMPLETE THEORY OF EVERYTHING...

This great theory is the ultimate knowledge of the universe! Shockingly, the five seemingly separate pieces of the universe: matter, space, time, energy, and the forces of nature are not five separate pieces at all: they are finally revealed for what they really are, manifestations of just one thing, one object! This one Thing → what we call "Space" now explains them all. It explains why they exist, how they work, and how they interact with each other.

> [Note: the word "space" has a bad reputation. When we hear it we think of a void, of nothingness. But this is a mistake because space is not a void; it is made of something and might even possess consciousness, and emotion? (Ch. 38)]

In the past, a number of individuals have falsely claimed that their theory about the construction of the universe is the Theory of Everything! However, when they are asked to use it to explain some of the great mysteries of the physical universe, they fail miserably! They have no idea how to explain such daunting, never explained mysteries as... *the unification of Newtonian Physics and Quantum Mechanics; the Force of Gravity; Quantum Gravity; Quantum Entanglement; the Constant of Fine Structure (the mystifying: dimensionless number of 1/137); Dark Energy; Dark Matter; the Pauli Exclusion Principle; or how single photons create Double Slit Interference Patterns; (and many...many more)*! [Note, all are now easily explained via this *Vortex Theory of Atomic Particles* and are in this book. Its significance is simply stunning, overwhelming!]

ON THE SHOULDERS OF GIANTS...

I must acknowledge too, that like Newton; if I had seen further than others it is because I have stood upon the shoulders of Giants: Albert Michelson, Edward Morley, Hendrik Lorentz, & George FitzGerald.

More than a 130 years ago, an elaborate succession of experiments was begun in the early 1880's and ended in 1887 by Michelson and Morley to detect the presence of the Aether: the substance it was theorized space was made of. They reasoned that as the Earth plowed through the Aether, space would be blown to either side of it like the wake of a ship plowing through the ocean. But their apparatus failed to find the wake! But it did discover some other strange results that later have come to be called: length shrinkage; and time dilation.

Enter Hendrik Lorentz and George FitzGerald. These two great scientists working separately from each other, and living in different countries, both theorized that the results of the experiment could be explained if matter in the direction of velocity shrank; and, Lorentz also added that if time itself slowed down [this was before Einstein's theory of Relativity] it would explain why no difference was seen in the data from Winter to Summer when the earth speeds up or slows down as it orbits the Sun. But their explanations for length shrinkage and time dilation were discounted by that era's scientists as being a contrived "Ad Hoc" explanation: and rejected.

In looking around for another theory that might explain the unexpected results of the experiment, the search led to the rise of Einstein's theory of Relativity. He satisfied a few of the scientists of that era by proposing [offering no proof!] that space was made of nothing [hence no Aether wind]; that time slows down at high velocities; and that the length of objects shrink. Both phenomena he claimed were being created by a fourth dimension he called "Space-time". He called it Space-time because he theorized that it supposedly possessed both space and time characteristics. Note: it must again be understood that he offered no proof whatsoever! It was pure supposition, and speculation on his part! And there has been absolutely *no proof* whatsoever by anybody that this imaginary fourth dimension of "space-time" causes length shrinkage and time dilation.

However, it is interesting to note that the Michelson Morley Experiment did indeed offer *a proof* that nobody of that era or this one realized. What they proved was that the "Relationship" the scientists of that era believed existed between space and matter was a mistake!

This relationship said space was made of something and matter was made of solids. Solids that would in turn create a wake as the Earth plowed through space. ***It was this idea of a "wake" that the Michelson Morley Experiment really disproved***! There was no wake because matter was not made of solids. They did not know it, but matter exists in a manner unimagined by the scientists of that era, or today!

The thesis that the sub-atomic particles all matter is created out of are made of three dimensional holes in space creates the exact results that the Michelson Morley experiment discovered! Furthermore, there is no Aether Wind created as this massive collection of three dimensional holes plows through space, because space reconfigures itself around these holes as they move, and no Aether wind is produced!

> The construction of matter is the key to understanding how the universe is really constructed. Contrary to popular opinion, space is not made of "nothing"! Instead it is made of something, and it is matter that is made of nothing! A shocking reversal of ideas!!!

Chapter 4
It Is Space That Is Made of Something
and Matter That Is Made of Nothing!

The following is from part of a conversation that took place between me and a professor of physics at Florida Atlantic University in West Palm Beach Florida in the Summer of 2005…

"…What!… You say space is made of something; and that matter is made of nothing! That's crazy!"

"Didn't the scientists in the late 1880's reveal that the Aether [the substance they believed space was made of] didn't exist? Didn't they reveal that to allow high frequency energy to pass through it, space would have to possess the density of steel?"

"Even more important, didn't the famous experiment conducted by the two greatest scientists of that era → Michelson and Morley at Case Western Reserve University in Cleveland Ohio to find the Aether → spend many disappointing years trying to find the Aether only to fail? Didn't they continually develop more and more sophisticated apparatus, for more and more elaborate experiments, only to finally conclude that they could not find it? And just before they finally quit, didn't they make one last desperate attempt to try to eliminate unwanted vibrations disrupting their equipment by placing their apparatus upon a two ton sandstone slab of rock and float it in a massive pool of mercury: and still they failed?"

"Sadly…for those who once believed whole-heartedly in the Aether Theory such as Maxwell, and Faraday whose life works and formulas were based upon the existence of the Aether…isn't it good that they are now dead, so they don't have to listen to all this nonsense anymore? Isn't the Aether a dead subject too…?"

I replied with the authority of the PhD I just had been awarded by the Russian Ministry Education for the mathematical thesis I presented on this subject: "No it is not a dead subject...Not quite…Not quite at all…!"

I went on to say, "For more than 100 years, we have all been misinformed, and misled; because the *whole story* has NOT been told! Forgotten is one equally important fact: that in this by-gone-era, not only did scientists believe that space was made of something; they also believed [and is left out of most science text-books] that matter was made out of Solid objects. [The particles that make up atoms – electrons, protons, and neutrons – had not been discovered yet!] Consequently, what was really disproved was this mistaken 'RELATIONSHIP' between space and matter! Not the Aether!!!

Not at all!!! Not one bit!!!"

What these Great scientists of yesteryear [Michelson was a Nobel Prize winner; Morley was famous for measuring the atomic weight of the oxygen atom] never considered was that another relationship might exist between matter and space that they knew nothing about! If they could have leaped into the future and investigated the findings of the *Vortex Theory Thesis*, they would find out that atoms of matter are created out of protons, electrons, and neutrons that are really three dimensional holes in space. If they knew this, they might have then discovered the real reason for the length shrinkage and time dilation effects. But it was not their fault because these three particles were not discovered until long after Michelson and Morley had conducted their famous experiment in 1887: the electron by J. J. Thomson in 1897; the proton by Ernest Rutherford in 1920; and the neutron in 1932 by James Chadwick.

The discovery that space was indeed made of something, and that these three "particles" were really holes in space was discovered by the *Vortex Theory of Atomic Particles* in 1989 and first revealed in Book 1 of this series; and proven in the *End of Time* PhD thesis. This brand new relationship between space and matter whose mathematics reinvestigated the results of Michelson and Morley's great experiment, proved that space was made of something, and matter was made of nothing at all! Shockingly, the "particles" that all Matter is made out of were discovered to be simply three dimensional holes existing upon the surface of fourth dimensional space! A simply astonishing discovery![3]

---

[3] This mathematical End of Time thesis was presented, and published in the Appendix of Book 1. It was put into the appendix because it is known that most people do not care to read mathematical descriptions [I, am one of them!]. Nevertheless, this mathematical description of the results of the Michelson Morley Experiment proved not only that time, "time itself" does not exist, but also that space is indeed made of something, and matter was made of nothing. Particles of matter are really three dimensional holes in space!! A triple conclusion for which a PhD in Nuclear Physics was given out in 2005 by the distinguished Russian Ministry of Education: and backed by the Russian Government!

The PhD Thesis was later published in 2012 by St. Petersburg State University's Branch of the Russian Academy of Sciences…

See reference 8. on page 166

> Einstein's mistaken proposal that space is made of nothing spawned another egregious mistake: "Particle Logic." Since space was supposedly made of nothing, it was then reasoned that everything that exists in the universe has to be made of particles; a great mistake that has lasted for over a hundred years!

Chapter 5
Einstein's Mistake Spawned Another Equally Egregious Mistake: "Particle Logic"

Before we can begin to explain the answers to the great mysteries of the universe such as Dark Energy and Dark Mass in the coming chapters, an egregious mistake in science must be exposed…

…today's incorrect scientific knowledge, and misunderstood subatomic phenomenon responsible for our erroneous thinking and reasoning processes can be traced to two major errors: the mistaken belief in Einstein's premise that space is made of nothing; and its equally disastrous consequences → Particle Logic!

Particle Logic was a mistake subconsciously developed by scientists reasoning that if Einstein's proposal is true, if space is indeed made of nothing, then everything that exists in the physical universe HAS to be made out of "particles". A false scientific reasoning process I have labeled: "Particle Logic"! A mistaken way of thinking and reasoning that has led scientists astray for the past one hundred years! A mistake that must be corrected now → and reversed now, before it becomes like the incorrect assumption of Ptolemy's Wheels in the Sky and lasts for 1300 years!

Unfortunately, the greatest scientists and thinkers of today's world use this false Particle Logic as a way of reasoning to try to explain complex sub-atomic phenomenon such as mass: specifically the Higgs Boson Particle!

THE MISTAKE OF THE HIGGS BOSON PARTICLE!!!

Mass is not created by a particle! **Mass is created by the resistance of the spherical surfaces of the subatomic holes matter is constructed out of, and is being distorted.** So the particle CERN has supposedly "discovered" is unknown! Whatever it is, it is not a particle explaining mass!

A Higgs Boson Particle explaining mass or any particle attempting to explain mass does not exist! So just what did they find at CERN? Your guess is as good as mine? But what is true is this: "you cannot find something that does not exist!"

Consequently, this "mistake" must be cleared up! At stake are the educations of millions of students studying science and engineering. It is outrageous to realize they are paying a lot of money to be taught a completely false narrative!

PART II
SPACE CREATES EVERYTHING IN THE UNIVERSE!

> The thesis that one thing, one object creates everything in the universe begins with space. To explain how space is created, we use the parable of the blind men and the elephant.

Chapter 6
What Is Space Made of ?

We are about to explain the mysteries of the universe; but before we do, space must be revealed for what it really is. Unfortunately, the following is a sample of what we have to deal with when talking to conventional scientists…

Here is a part of a grimly mocking, cynical & sardonic conversation I had in the winter of 2006 at a university in South Florida; with a rude professor of physics whose expertise was the String Theory…

"So"…if what you say is true…if space is made of something…then what is it made of?"

After a long pause of thinking about how to answer this question…I said, "Perhaps we will never know, because there is nothing to compare it to. It is a unique substance; unlike anything we have ever encountered before in the universe. It is an answer that involves both Science and Religion!"

"Both?"

"Yes…Both?"

"Let me remind you of something you need to be aware of…at this university we do not mix science and religion! Is that clear? So what do you have to say now?"

"Well…scientifically, what we can do is explain what space is capable of doing: it can flow, bend, deform, expand & contract, become denser or less dense, and rip open → creating holes in its surface that extend into its interior; which appears to be made of at least four more higher dimensions: making a total of at least seven."

"How about this Religion bit…?"

"Well, perhaps most shocking of all… space *could possibly* possess consciousness?"

"Really…Consciousness you say???"

"Yes!"

"This is not a fact, but a conclusion based upon the evidence of how space creates our own consciousness!" [A conclusion each reader will come to naturally on their own at the end of this book after they have seen all of the evidence presented!"]

"Nonsense!" he replied. Then he said with a wave of the back of his hand as if to callously dismiss me, "I have a class to teach. Goodbye!" [Who starts a class at half past twelve in the afternoon?]

I started to leave, when hanging on his hat rack, I spied a black canvas shoulder bag he had received from attending a String Theory Conference, and this gave me an idea. So I turned to him and said, "One last question… [and without waiting for him to reply I continued]…My thesis for which I received a PhD in Nuclear Physics from the Russian Ministry of Education; and certified by over 33 physicists, 18 of which were chairmen of their departments, and 15 other tenured professors from universities all over Russia; reveals not only that space is made of something, and that time does not exist; but also, there is no fourth dimension of Space-time. And if there is no fourth dimension of Space-time, can the String Theory still explain the Length Shrinkage and Time Dilation effects that take place at near light velocities?"

After a slight pause, he stared at me, evaluating me, and realized that not only was I not a gullible naive student he could act superior to, and give whatever answer he chose; but having a PhD in Nuclear Physics the same as him, and probably noting the confidence with which I asked the question, he probably realized I already knew the answer [which I did]…he reluctantly had to say…"No."

[Which is the correct answer; because without space-time, the string theory *cannot* explain length shrinkage and time dilation. And any theory that cannot explain both of these phenomenon is null and void!!!]

With that I turned my back on him and walked out of his office, leaving him sitting there thinking about his answer! So much for talking to the establishment and self-absorbed professors with inflated opinions of themselves. I felt sorry for his students, who have to tolerate someone who teaches but does not listen!

[Then again; "Accentuating The Positive as it says in the song"…I would like to say that this man did have a fastidiously orderly nice office; with a well-organized nice bookcase; with everything nicely in place; in his nice office on the third floor of the nicely well kept ……. Physics Building; with a nice, good view to boot…

…it was nice! ☺]

THE PARABLE OF THE BLIND MEN AND THE ELEPHANT

To better help us to understand what space is capable of doing, we return our attention to the ancient parable of the elephant and the blind men. If you recall, or if you never heard about it, one version of it goes something like this: six blind men heard that an unknown animal called an elephant had entered their village in India. Curious to find out what the elephant was, the head blind man said, "…let us go forth and investigate this strange creature." (And so they did!)

As they approached the elephant, the head blind man said, "…since we cannot see it we must investigate the elephant using touch."

So as they gathered around the elephant, the first blind man touched the trunk and said, "The elephant is like a big snake;" another who touched its ear said, "Your wrong! The elephant is like a big fan;" the third touched its leg and said, "Your wrong! It is like a tree trunk;" the blind man who touched its side said, "Your both wrong! The elephant is like a wall;" the blind man who touched its tail said, "None of you are right! The elephant is like a rope." While the last blind man who touched its tusk said, "None of what any of you say is true! The elephant is like a spear."

Although this is a quaint little proverb mainly for children, it has an interesting moral: that many people tend to claim they have discovered absolute truths based upon limited personal experiences; while at the same time, ignoring other people's own different individual limited experiences that might also be true: causing them to make erroneous conclusions. And so it is with "Space."

Many scientists at the beginning of the 20th Century believed that the Michelson Morley Experiment had failed to find the Aether. To them, because this lone experiment failed, it revealed that there was no Aether. But just like the parable of the blind men and the elephant, this was not true.

MICHELSON & MORLEY WERE REALLY LOOKING FOR THE AETHER WIND

Again, it is very important to know that what the Michelson Morley Experiment was actually looking for was the "Aether Wind." Consequently, because the Michelson Morley Experiment failed to find the wind, it was *assumed* by Mr. Einstein that the Aether did not exist!

However, what was really disproved was the "*Relationship*" between the Aether and Matter. Then, Matter [atoms] was believed by some to be made of solids that were condensations of space, much like ice in water. This idea was a 2500 year old theory proposed by two ancient Greeks: Leucippus and Democritus. Unfortunately, what nobody realized then was that another relationship between these two different pieces of the universe might actually exist.

Instead of matter being made of solids like rocks, the atomic particles that construct atoms: protons, electrons, and neutrons as tiny three dimensional holes in space; create a different interaction with space! These holes in space do not push the substance space is made of to either side as matter moves, but rather, space reconfigures itself around these holes and no Aether Wind is created!

It is too bad that Michelson and Morley did not know about protons, electrons, and neutrons. If they did, they would have realized that another relationship might exist that could explain their unexpected results. These were brilliant men. Michelson was no ordinary scientist, but was recognized as the greatest experimenter of his time: he actually performed an experiment to measure the diameter of the star Betelgeuse: a simply fantastic achievement! As mentioned before, Morley measured the mass of the Oxygen Atom: another great achievement for that era. Also, it must be said, that Michelson continued to believe in the Aether until his death in 1931. He never believed in Einstein's mistaken postulate that space was made of nothing![4] What a tragedy, if this great man only possessed options!

---

[4] 1984; Dorothy Michelson Livingston (Michelson's daughter); One Pass Productions; Cinema Guild; The Master of Light: A Biography of Albert A. Michelson. University of Chicago Press.

> The knowledge presented here about "Time" comes directly from the thesis that was awarded a PhD [called *A Candidate of Science and Engineering* in Russia] in Nuclear Physics in 2005 by the Russian Ministry of Education.
>
> Although the original mathematics of the thesis was presented in the appendix of Book 1, and can be found there, it is not presented in the following explanations because it is acknowledged that most people do not like to have to endure abstract mathematics.
>
> It is enough to state that this mathematics revealing time does not exist as a fundamental principle of the universe was confirmed and endorsed by none other than the powerful Russian Ministry of Education that awards all higher degrees in Russia. In 2012, because of its revolutionary nature, this thesis was again confirmed and approved before being published by the St. Petersburg State University's distinguished 10 Professor peer review governing board for articles in the University's Journal Branch of the Russian Academy of Sciences.

Chapter 7
Time Does NOT Exist; Hence, Space Does Not Possess Time Characteristics!!!

To explain the great mysteries of the universe, the mistake called "Time", time itself must be addressed.

The following outrageous conversation took place in June 1985, at the home of a former director of the Los Alamos National Labs in New Mexico…

…"What the hell do you think you are talking about…don't you tell me time doesn't exist! You're no scientist!"

"Let me remind you that you are nothing but a swimming pool builder and not a physicist!"

"I am a physicist and spent a lot of years becoming one."

"If you are going to tell me something about swimming pool construction, I might write it down because some day I might want to build a pool; but don't you ever tell me anything about science because you don't know anything about science!"

The tirade continued…

"What is that on the wall?"

[I replied], "A clock?"

"Yes a clock! And what the hell is it keeping track of?"

[I said], "It is not keeping track of time. It is keeping track of the motion of the Earth; and was originally developed as a navigational device for sailors to determine longitude!"

"No! It is keeping track of time!" And with that he stomped out of his own house, leaving me sitting alone at his kitchen table!

The above is a true conversation that took place many years ago between me and a renowned scientist at the Los Alamos National Labs in New Mexico. This was before I obtained a PhD in Nuclear Physics, the same degree this man had.

During this conversation, I unfortunately learned a disgusting truth some scientists believe in: that in America, only a physicist can talk to a physicist about physics. The rest of us are either considered uneducated and do not know any better; or too stupid and cannot know any better. It's true!

> Only in the Russian Educational System do professors listen to the opinions of their students. Because *before* a student can apply for a PhD he has to have written *five* peer reviewed scientific papers and have them published in worldwide, international conferences on science. Consequently, their professors and advisors have to listen and evaluate what they have to say, and their mathematical analysis before a paper they are about to write will be capable of passing peer review.

Unfortunately, during this conversation, I learned one of the cruel truths of mankind. I learned from this conversation with this former director, that when talking to others, you will create either an emotional reaction or an intellectual reaction. The first reaction results from the perception of an attack upon one's beliefs, which is then perceived as an attack on oneself, causing a need to [either rightly or wrongly] defend themselves. The second reaction is a result of hearing knowledge needing reflection and contemplation and is not perceived to be a threat to ones beliefs or oneself. [Example: a negative emotional response is instantly created..."Your kid is ugly?" Verses the positive creation of the intellectual reaction..."What is the distance between New York and Chicago?"]

Although many scientists consider themselves above emotional reactions and like to consider themselves intellectually above the "fight or flight" response, but this is not true. There are concepts they have lived with all of their lives that they accept and believe to be true. And when someone challenges them, they are offended, and the emotional reaction suddenly rears its ugly head!

This was blatantly obvious many years ago when Albert Einstein and Niels Bohr got into an intellectual disagreement and argument that ended in an emotional confrontation at the *Fifth Solvay International Conference* on Electrons and Photons in 1927. Einstein did not like Bohr because Bohr was an advocate of Quantum Mechanics which proposes that the universe is created by random actions and ruled by the laws of probability. In response to which Einstein made his famous statement: "…God does not play dice!"

He made this statement because Quantum Mechanics seems to suggest that "order is created out of random disorder"; a belief that offended Einstein's belief in God: a God that created the universe using the principles of harmony and order, and not disorder. Which brings us to today's problem: the concept of time…

…we have lived with the concept of time for so long, that to hear it doesn't exist creates an emotional response that makes one want to immediately reject this idea; unlike the casual and intellectual response that allows us to calmly investigate evidence and eventually make an unemotional decision to accept or reject.

To help the process pass from the emotional response to the intellectual response, we must remember that we did not invent time. The concept of time is so old it predates writing…

…[a quote from book 1: *The End of the Concept of Time*]… "The early civilizations documenting the concept of time were chronicling a concept created much earlier than the creation of writing. Note: this idea is not speculation. In 1993, it was reported in *SMITHSONIAN TIMELINES OF THE ANCIENT WORLD* [see Reference 1) on page 167], that a carved bone discovered in the Grotto du Tai in France appeared to be a solar calendar dating back to 10,000 BC."

Hence it is kind of a shocker to realize that we, as modern human beings, are still using concepts to think with and reason with that were originally conceived of in the ignorant childlike minds of our caveman ancestors living thousands of years ago! Although millennia ago, we discarded the clubs they used as weapons, shockingly, we still think and reason with the same logic and thought processes they developed! [How does the besmirched saying go, "Even a caveman can do it!" Well, I hate to say it, but it appears… the cavemen have had the last laugh!]

THIS LITTLE OBSERVATION HELPS TOO…

…if *every* motion of *everything* in the universe suddenly stopped and then started back up again, is there any way to tell for how long it was stopped?

…after much contemplation, the answer is NO! Every natural or manmade way used to keep track of time is in motion, [even tree rings are created by chemical reactions taking place at fixed rates]. In fact, because the motions of all of the atoms in our bodies would have stopped too, including the electrons flowing through the neurons in our brains responsible for our thought processes, we would never even know it had stopped!

Although this is merely a philosophical question, for science, it possesses profound implications. For thousands of years man has believed in the concept of time; but if time stops when motion stops, has anybody ever realized that time might not be a real "piece" of the universe?

One of the great discoveries of the Vortex Theory of Atomic Particles is that one of the five pieces the universe is supposedly constructed out of → "time", does not exist! Instead, what the Vortex Theory Thesis discovered, and mathematically proved was what we call "time" is really nothing more than a function of motion, a phenomenon created by motion; a shadow of motion; and as such, it does not exist as a fundamental principle of the universe.

This discovery possesses shocking implications for the construction of space. Because if time does not exist as a fundamental principle of the universe, it is not real! And if "time" is not real, then Einstein's 4th dimension of "space-time" does not exist either. [Note: a 4th dimension of space does exist but possesses no "time" characteristics.] This reveals a stunning revelation: that space does not exist as 20th Century scientists believed it did!

This creates a problem for 21st Century Scientists, because if time does not exist then Mr. Einstein's Space-time does not exist either.

And…if Space-time does not exist, space does not possess any "time" characteristics. Consequently, something else has to be responsible for the length shrinkage and time dilation effects that take place at near light velocities. But even more important, if "space-time" doesn't exist, then Einstein's vision of the construction of space is also a mistake. And since the construction of space also defines how matter is constructed, the 20th Century Scientific explanation of the construction of particles is a mistake too. Matter cannot be made of solids; because if indeed it was, the Michelson Morley Experiment would have detected the presence of an Aether wind caused by the Earth's motion through space. Nor can space be made of nothing because if it was, it can no longer create and explain length shrinkage and time dilation! So how is matter constructed?

Many years ago, while contemplating of all of the above, it was realized that a revolutionary vision of space and matter must exist that can explain both length shrinkage and time dilation without the use of "time", or space made of nothing. This revolutionary discovery was made and eventually led to what has since come to be called *"The Vortex Theory of Atomic Particles"*. This revolutionary theory was responsible for discovering a brand new relationship not only between matter and space; but also between energy and the forces of nature…

…and, in the process, this new relationship began to explain not only all of the mysterious phenomenon we see in the universe; but also the rediscovery of how *everything* we know appears to exist: such as space…

> The beginning of the analysis of space begins with the realization that all surfaces are one dimension smaller than the volume they enclose; just as all holes in a volume are one dimension smaller than the volume they penetrate.

Chapter 8
It All Begins With Three Dimensional Holes in Space!

Years ago, when I began this lifelong odyssey that ended up with my having to re-examine and revamp the scientific vision and knowledge of the universe and prove the PhD thesis presented in Book 1; I realized I needed to answer a critical, very important scientific question that has profound religious implications: does a fourth dimension of space actually exist? Was there any proof? And was there any precedent upon which to justify this truly extraordinary and revolutionary conclusion?

The answer was yes!

From my previous studies of space and its dimensions, I realized that in three dimensional space, there could only be two dimensional holes. All pipe ends, doors, windows, cave openings, etc. are two dimensional holes. Furthermore, the tops and bottoms of tornadoes, hurricanes and whirlpools are two dimensional openings allowing passage into the third dimension.

The mental picture of these two dimensional openings allowing passage into the third dimension is the analogy allowing us to realize all holes are one dimension smaller than the number of dimensions that are present.

This relationship occurs because each surface is one dimension smaller than the volume it envelops. For example, the surface of a three dimensional volume such as a bowling ball is two dimensional, while the surface of a fourth dimensional volume would be three dimensional. With these ideas in mind, it is easy to see that to enter into the interior of a volume, its surface must first be penetrated. Because the surface is always one dimension smaller than the volume, the hole is always one dimension smaller than the volume it enters into, or out of. Hence the openings for the thumb and finger holes on the bowling ball are two dimensional.

Therefore, if fourth dimensional space does indeed exist, even though it cannot be seen, one characteristic indicative of its presence will be the existence of three dimensional holes somewhere in our universe! Just like two dimensional holes allow passage into the third dimension, three dimensional spherical shaped holes allow passage into the fourth dimension.

Note: if fifth-dimensional space exists, there will be additional fourth dimensional holes within these three dimensional holes, and so on and so forth depending upon how many dimensions exist, and how many dimensions are penetrated. (An important observation that will become obvious later on when the particles called quarks are explained.)

So the question is this…do these three dimensional holes actually exist?

It took a while to find them, but when they were found, it was a spectacular yes! A resounding Yes!!!

They are all around us, and are us! They are discovered to be the protons, electrons, and neutrons that all of the matter in the universe is made of!

> Note: we know protons and electrons are spherical because their electrostatic charges are spherical. If they were of any other shape, their electrostatic fields would be distorted.

To better understand the relationships that exist between a volume, its surface, and holes that penetrate into its volume; the following drawings are made for a quick observation right now; later they will take on a great significance when quarks are explained in the next chapter...

A two dimensional volume, [envision a piece of paper]; the (2d) volume has a one dimensional (1d) surface [envision the edge of the piece of paper]. A hole penetrating into its interior has a one dimensional (1d) opening cut into its (1d) surface...

Figure 8.1   **Figure 8.2**

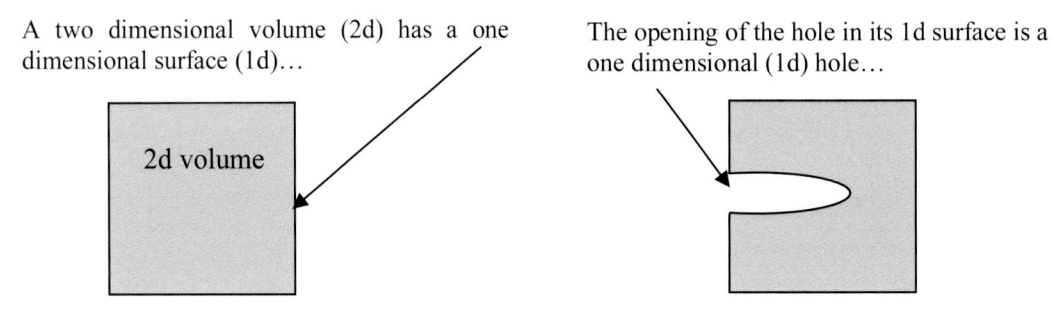

A two dimensional volume (2d) has a one dimensional surface (1d)...

The opening of the hole in its 1d surface is a one dimensional (1d) hole...

A three dimensional volume has a (2d) surface. A hole penetrating into its interior has a (2d) opening...

Figure 8.3   **Figure 8.4**

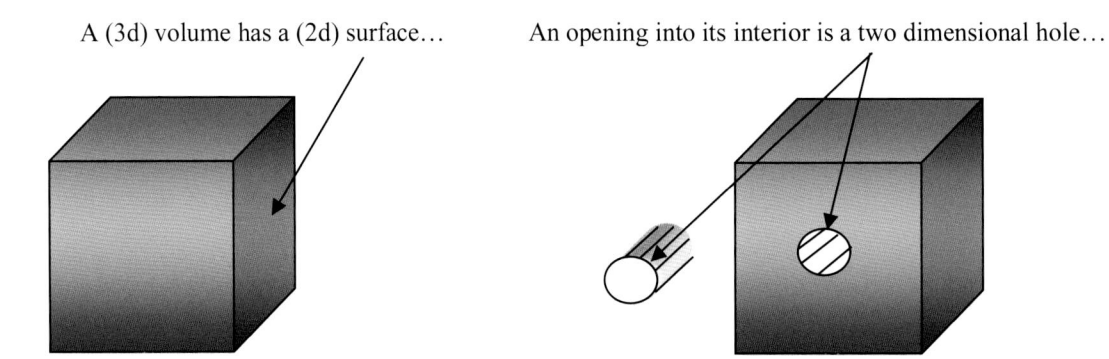

A (3d) volume has a (2d) surface...

An opening into its interior is a two dimensional hole...

A fourth dimensional volume is impossible to draw. It has a (3d) surface that is also impossible to draw. However, a hole penetrating into its interior has a three dimensional (3d) opening…

Figure 8.5   **Figure 8.6**

A (4d) volume has a (3d) surface (?)   An opening into its interior is a (3d) hole!

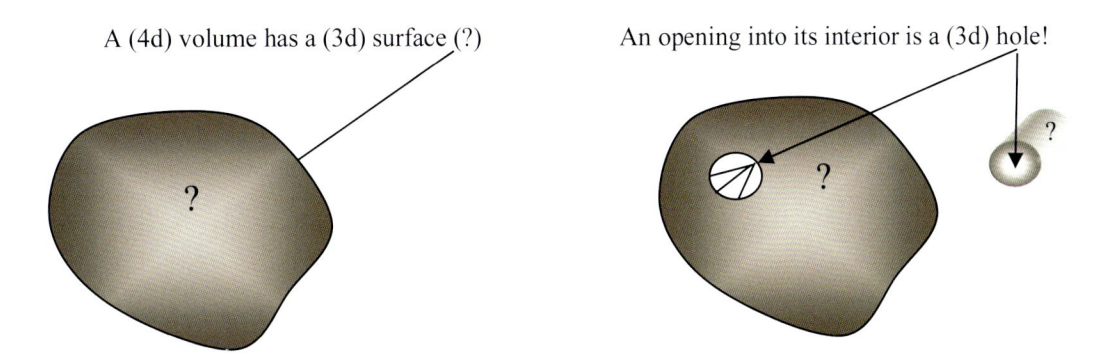

A fifth dimensional volume is impossible to draw. It has a (4d) surface that is also impossible to draw. However, a hole penetrating into its interior has a four dimensional (4d) opening…

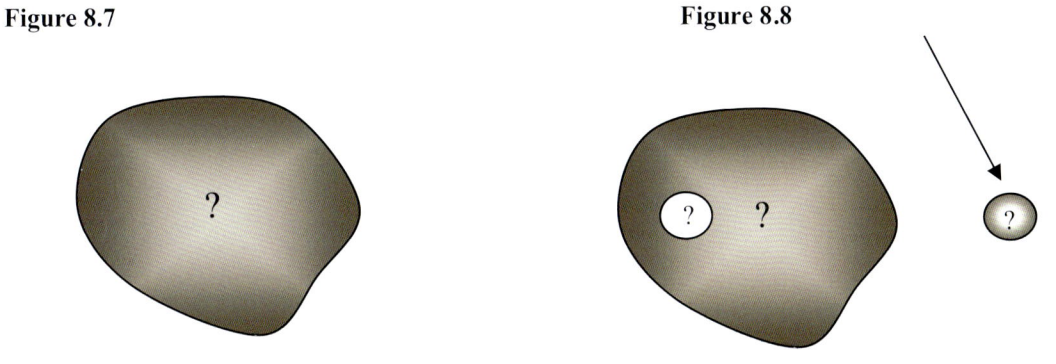

Figure 8.7   **Figure 8.8**

A sixth dimensional volume is impossible to draw. It has a (5d) surface that is also impossible to draw. However, a hole penetrating into its interior has a five dimensional (5d) opening…

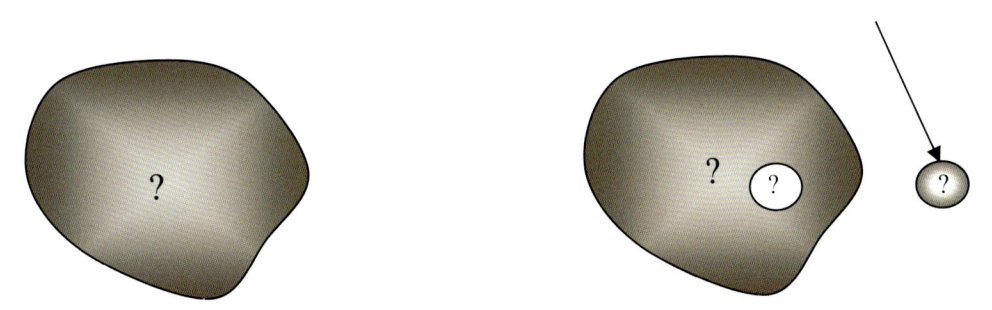

Figure 8.9   **Figure 8.10**

A seventh dimensional volume is impossible to draw. It has a (6d) surface that is also impossible to draw. However, a hole penetrating into its interior has a six dimensional (6d) opening…

Figure 8.11

Figure 8.12

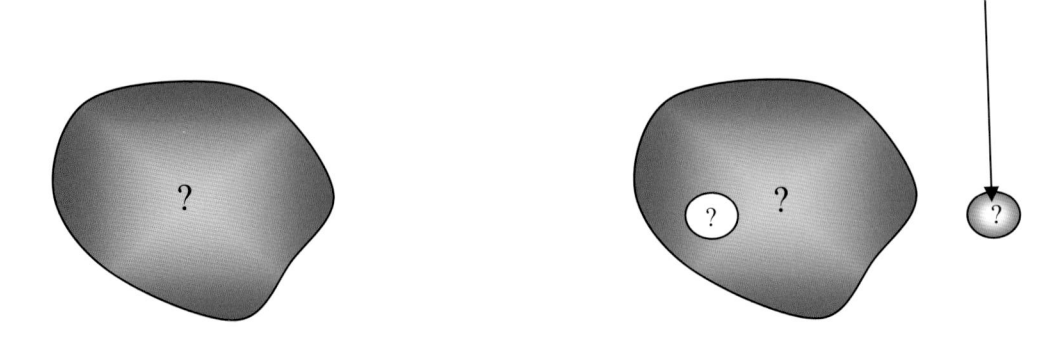

> Like the story of the blind men and the elephant… we can begin our inspection of space by first stating that "Space is like a fabric because it can rip open creating three dimensional holes in its surface that become the sub-atomic particles of matter…

Chapter 9
Space Is Like a Fabric That Can Rip Open Creating 3d Holes in Its Surface

According to the *End of Time* thesis space is made of something. This is *not* a return to the old Aether Theory where it was mistakenly believed space was made of something and matter was made out of condensations of space like ice in water. Such a scenario creates an Aether wind as matter moves through space: and the Aether wind was eliminated by the Michelson Morley Experiment. Hence, if space is made of something matter cannot be made of solids. Matter has to be constructed in some other manner.

This is where this Vortex Theory succeeds most remarkably! Because according to this theory, "particles" of matter such as protons and electrons are not solid objects at all. Instead they are hypothesized to be three dimensional holes existing upon the surface of fourth dimensional space. Consequently, as these holes move through three dimensional (3d) space, the surrounding 3d space reconfigures around them and no Aether wind is created.

The Particle and Wave Effects are suddenly and beautifully explained by realizing the particle effects of matter are created by the 3d holes in space [we will discuss quarks later]; while the wave effects of matter are created by regions of less dense or dense regions of space as 3d space flows into or out of the hole. Protons are surrounded by regions of less dense space as 3d space is pulled into its 3d hole. Electrons are surrounded by regions of denser space as 3d space flows out of its 3d hole. Pushing the surrounding space outward, making it denser.

Figure 9.1

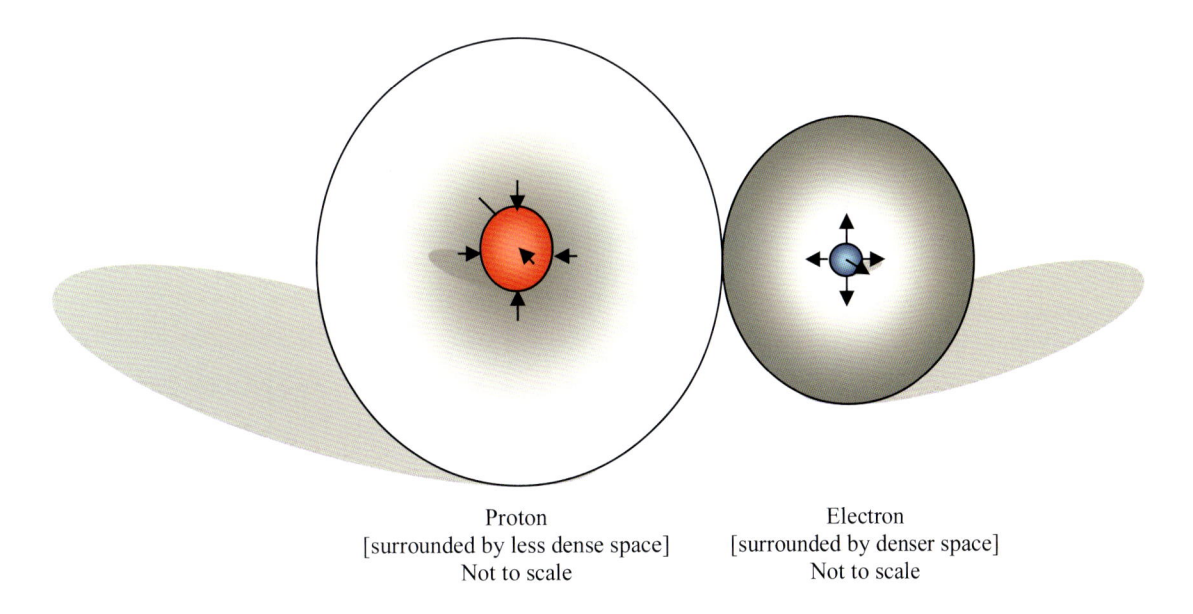

Proton
[surrounded by less dense space]
Not to scale

Electron
[surrounded by denser space]
Not to scale

The proton is a 3d hole bent *into* the surface of 4d space; 3d space flows into the proton creating its electrostatic charge [note: contrary to present belief, quarks DO NOT create its charge: quarks will be explained later]. The electron is a 3d hole bent *out* of 4d space; 3d space flows out of the electron creating its electrostatic charge.

Because space is pulled into the proton, it is surrounded by a region of less dense space. Likewise, because space flows out of the electron, it pushes the surrounding space outward creating a region of denser space. These important concepts will be elaborated upon later when Gravity and Anti-gravity are explained. For now it must be realized that as either "particle" moves through space, space reconfigures around the particle, changing its density as the particle passes through. This will be detailed later on when the explanation for the science called Quantum Mechanics is explained.

Also, it was thought a long time ago that space flows into one particle and out of the other. But where is it going? This led to the incredible hypothesis in the PhD thesis that these three dimensional holes are really the ends of invisible 3d vortices going into and then out of 4d space, [see Figure 9.3] connecting the proton to an anti-proton, and the electron to a positron. Three dimensional space flows into the proton, through a 4d vortex and exits out of the anti-proton; likewise, 3d space flows into the positron, through a 4d vortex and out of the electron.

However, the next big question regarding proton electron pairs is how they became *entangled?* Particle collisions in linear accelerators reveal that when a proton is created, an anti-proton is created with it; and when an electron is created, its anti-particle [the positron] is also created with it too. So how do protons and electrons become entangled creating hydrogen atoms?

The previous question is answered by first understanding that the northern lights reveal that electrons and other negatively charged "particles" coming from the sun are attracted to the north pole of the earth [it is actually the south pole of the earthly magnet]. This reveals that the vortices between electrons and positrons can exist in long filaments at least 93 million miles long without breaking up. However, what happens if two higher dimensional vortices cross each other [Figure 9.2] in higher dimensional space?

From experimentation with electronic circuits it is known that electric current always seeks the shortest path between two different charges. Therefore, at first it was not unreasonable to assume that if the motions of two holes in three dimensional space caused their vortices to cross each other in higher dimensional space, the two vortices would break and reconnect themselves to the particular hole completing the shortest path back to the three dimensional surface.

However, it was later realized that the true mechanism via which this switch would occur would be the change in the cross sectional area of the vortex. Because the electron is smaller in diameter than the proton, yet possesses the same amount of electrostatic charge as the larger proton, both vortices are of different cross sectional areas but possess the same volume of flowing space. The cross-sectional area of the vortex between the electron and its anti-particle the positron is merely denser. Consequently, this increase in density would be seen by a longer vortex with a larger, less dense cross-sectional area, as an increase in the electrostatic charge. This apparent increase in the charge would cause the broken ends of the larger less dense cross-sectional vortex to reconnect the thinner but denser ends of the smaller cross-sectional vortex seen in Figure 9.2 below:

Figure 9.2

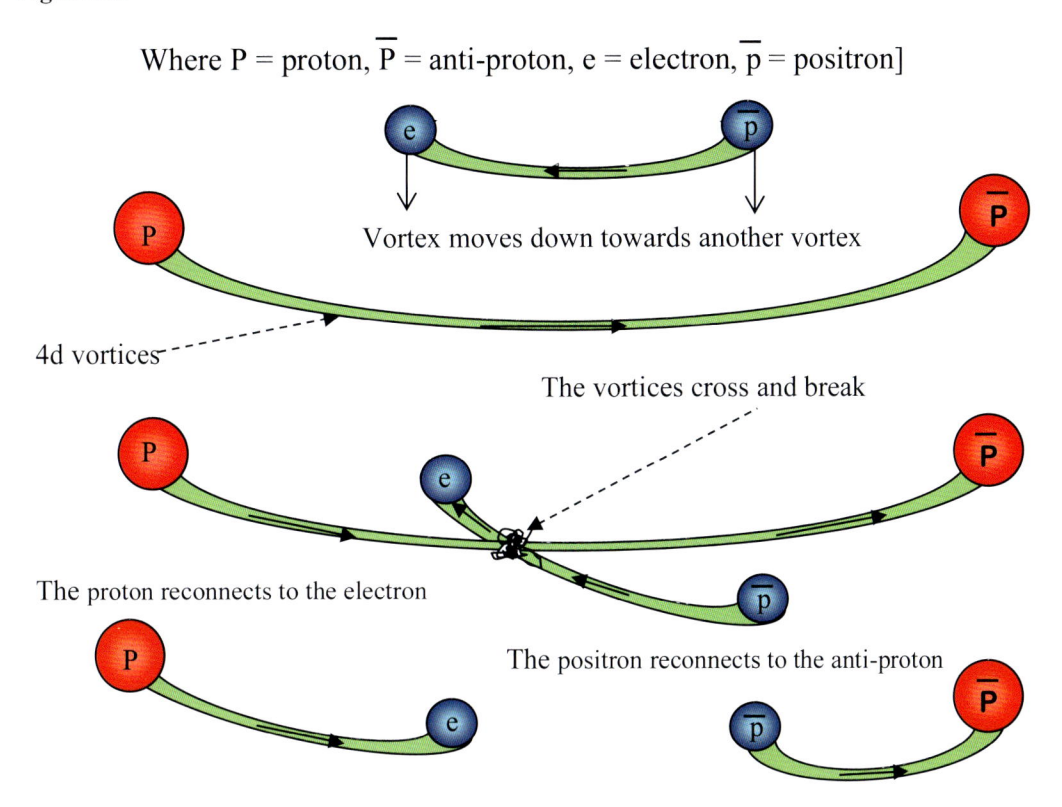

Where P = proton, $\overline{P}$ = anti-proton, e = electron, $\overline{p}$ = positron]

4d vortices

Vortex moves down towards another vortex

The vortices cross and break

The proton reconnects to the electron

The positron reconnects to the anti-proton

The proton and the electron are now connected by an invisible vortex of 3d space flowing from the proton to the electron in 4d space.

Figure 9.3

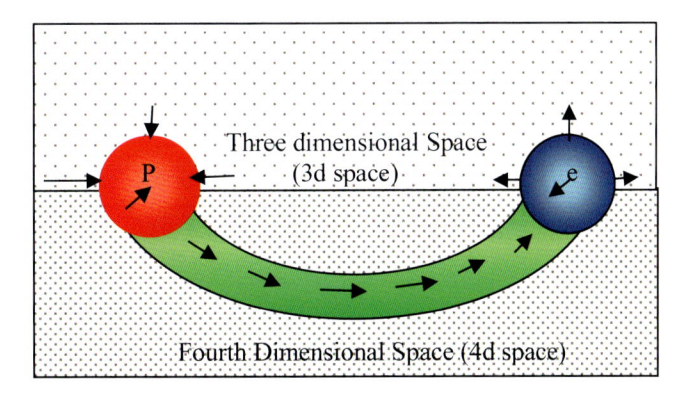

Two Vortices are created when a proton captures an electron eventually creating a *hydrogen atom* [Figure 9.4]. Above, in Figure 9.2, it can be seen that the vortices connecting them to the anti-proton and the positron break, reconnecting the anti-proton to the positron, and if close enough, such as in the early universe would eventually create an *anti-hydrogen atom* to be discussed later.

If the proton and the electron are close enough to each other and are pulled closer to each other by their electrostatic forces, a second vortex is created in 3d space when all of the space flowing out of the electron begins to flow into the proton: creating a situation seen below in Figure 9.4 below.

Space now flows from the proton, into 4d space through the 4d vortex, back into the electron; then out of the electron and back through 3d space into the proton: creating a circulating flow containing a fixed volume of space.

Figure 9.4

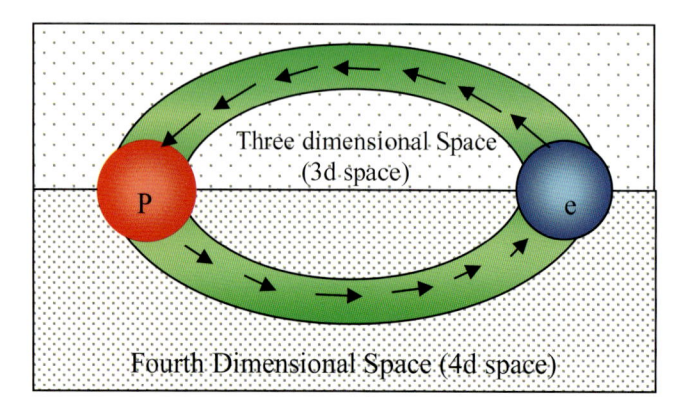

Note how the proton's and the electron's charges are now neutralized because all of the space flowing into and out of them are trapped in the three dimensional vortex.

Because the hydrogen atom is the simplest of all atoms, it is the example used in this analysis. However, since all protons and electrons *in* all atoms everywhere would also be connected by two vortices of flowing space. The principles introduced here apply to all other atoms in the universe as well.

When the circulating flow commences, both electrostatic charges are neutralized. The word "neutralized" was used because no flowing space escapes from the system.

An "Enclosed Loop" is created when the electromagnetic forces pulls these two holes together creating a hydrogen atom and the two vortices. This circulation creates an enclosed loop. The circulating flow begins, and the volume of flowing 3d space within the two vortices remains trapped, creating an unchanging constant volume.

This constant volume does not change. If it did, space would constantly be added or subtracted from the vortices causing all atoms to possess electrostatic charges. Hence, the volume of 3d space within this enclosed system remains the same no matter how fast the atom moves. In addition, because the radius of the atom is the length of one of the vortices, the radius of the atom now becomes a function of the maximum volume of space per unit of *measured* time that can flow from the electron to the proton or from the proton to the electron without backing up around either particle.

If space did back up, the speed of the vortices would drop below the speed of light and the electrostatic charges on protons and electrons would be variables instead of constants. And finally, although the electron and proton are of different sizes, this in no way affects our calculations. Because the charges on electrons and protons are of the same magnitude, the same volume of space flowing out of one particle flows into the other. In Book 2, *The Vortex Theory*, it is shown how the space surrounding the proton is less dense, caused by the surrounding 3d space being pulled into the proton's 3d hole. This region of less dense space allows its bigger though less dense surface to have the same volume of space flowing into it as that of the smaller diameter though denser space (caused by 3d space being pushing out of it) surrounding the hole we call the electron.

Figure 9.5

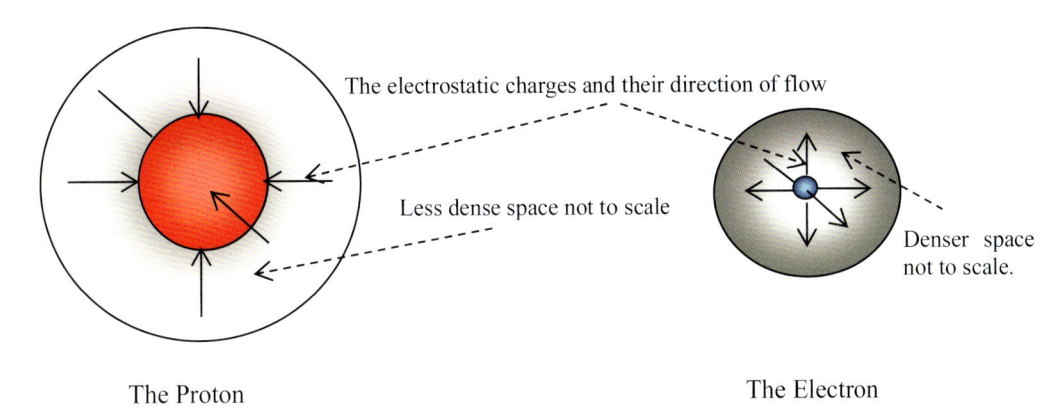

The electrostatic charges and their direction of flow

Less dense space not to scale

Denser space not to scale.

The Proton   The Electron

As the proton and electron are drawn together by the Coulomb forces, the formation of the "gradient of space" directly between them creates this "Corridor" of less dense, denser space flowing between them as seen in Figure 9.6 below. This vortex or "corridor" of space is less dense at the proton and denser at the electron; and creates a "cone shape:"

Figure 9.6 [Space flowing <u>from the proton to the electron in</u> **4d space**]

Note in the drawing below how even though the proton and the electron are of different sizes, the same amount of "lines" [3d space] flows into and out of each "particle" [hole].

Proton

Less dense space

Electron

Denser space

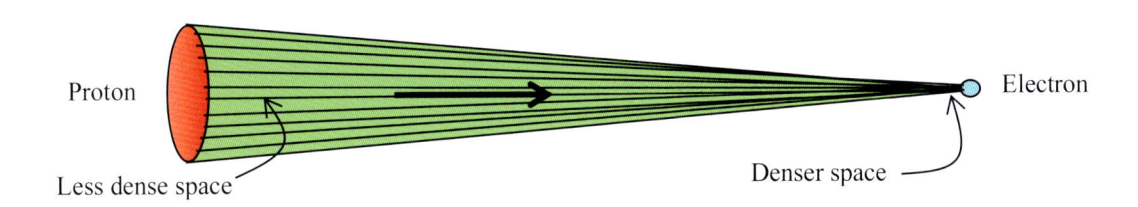

Figure 9.7 [Space flowing <u>from the electron to the proton in</u> **3d space**]

AGAIN: note in the drawing below how even though the proton and the electron are of different sizes, the same amount of "lines" [3d space] still flows into and out of each "particle" [hole].

Proton

Less dense space

Electron

Denser space

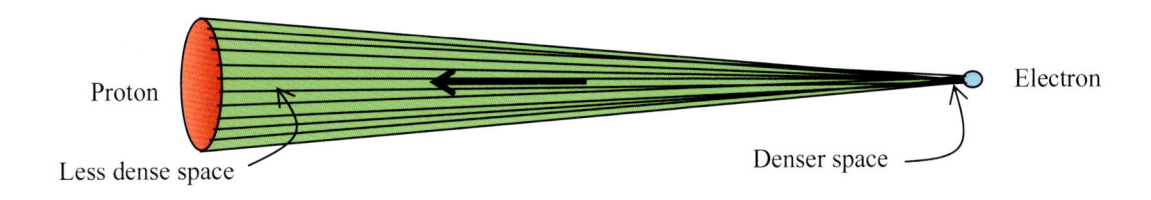

Note: ions are created when two atoms in a molecule are separated and a proton in one atom is connected to an electron in another atom via a 4d vortex that still flows between them:

Figure 9.8

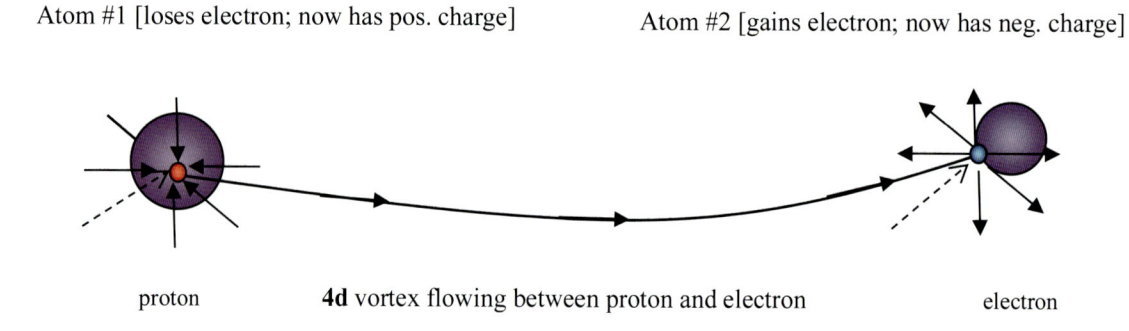

Atom #1 [loses electron; now has pos. charge]   Atom #2 [gains electron; now has neg. charge]

proton   **4d** vortex flowing between proton and electron   electron

Because the atom is so small, it appears as if the whole atom possesses the electrostatic charge, but this is not so. Three dimensional space is flowing into just one proton from Atom #1 giving it its positive charge; and out of the one electron lost from Atom #1, and captured by Atom #2 now giving it its negative charge. Note: because charged particles blasted out of the sun have vortices extending 93 million miles long, this vortex can be millions of miles long too.

PART III
THE PARTICLES OF NATURE

> The neutron represents a class of particles that seem to have no charge at all. But this is a mistake. The neutron is a hole within a hole! It is made out of a proton and an electron. Space flows from the electron to the proton, through 4d space, then instantly back into the electron whose size has increased dramatically, causing it to encircle the proton!

Chapter 10
The Secret of the Neutron

The subatomic discoveries we have come across so far can be described as bizarre, fantastic, and unbelievable. But the next discovery tops even these descriptions.

It is easy to see how electrons and protons are three dimensional holes. This discovery is revealed because their electrical charges are created by three dimensional space flowing into or out of them. But the neutron has no charge. This means no space is flowing into it or out of it. So how can it be a hole?

The answer is that the neutron is not just one hole; it is a hole within a hole! A simply fantastic concept!

Figure 10.1 As seen from outside the Neutron.   **Figure 10.2** Cross-section of the Neutron

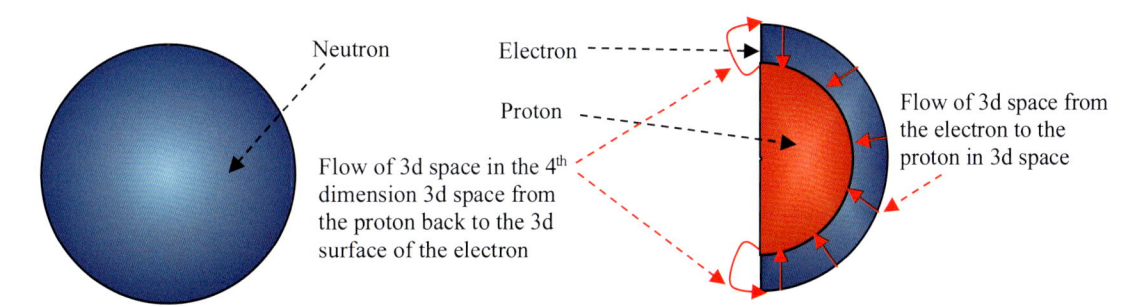

Neutron

Electron

Proton

Flow of 3d space in the 4th dimension 3d space from the proton back to the 3d surface of the electron

Flow of 3d space from the electron to the proton in 3d space

The neutron is created when an electron is shoved up against a proton and completely encircles it; or the vortex flowing out of a proton into higher dimensional space is hit by the right type of "particle", breaks, and completely encircles its three dimensional surface. Or, in a supernova blast of a dying star, the inner core of the star is crushed together, and all of the electrons are smashed into the protons creating the massive amount of neutrons that end up becoming a neutron star. [This supernova blast creates an enormous amount of light because all of the former vortices that once connected electrons to protons are disconnected from the two particles and are released all at once as a simply incredible amount of photons.]

Because the electron completely encircles the proton, the space flowing out of the electron is no longer flowing outward into the three dimensional space of our universe. Instead, its direction is reversed. It is now flowing inwards, directly toward the three dimensional hole (the proton) the electron is surrounding. This situation creates an enclosed loop.

The space flows out of the proton and into higher dimensional space; as soon as it does, it fans outward into a cone shape, is turned inside out, and instantly curls back upon itself creating a tight loop. This tight loop completes the return back into three dimensional space by flowing directly onto the surface of the encircling electron, forming a fourth dimensional torus - or donut. A simply fantastic shape!

Perhaps even more fantastic than the shape of the neutron is its speed of circulation. This circulation is taking place within the shortest distance imaginable, yet moving at the incredible velocity of 186,000 thousand miles a second.

The neutron has no charge because none of the space surrounding the neutron flows into it, or out of it.

The proton's surrounding less dense region of space encompasses the neutron giving it the same less dense region as other protons. The less dense space that usually surrounds the electron is not present because the space flowing out of the electron flows directly into the proton.

Figure 10.3

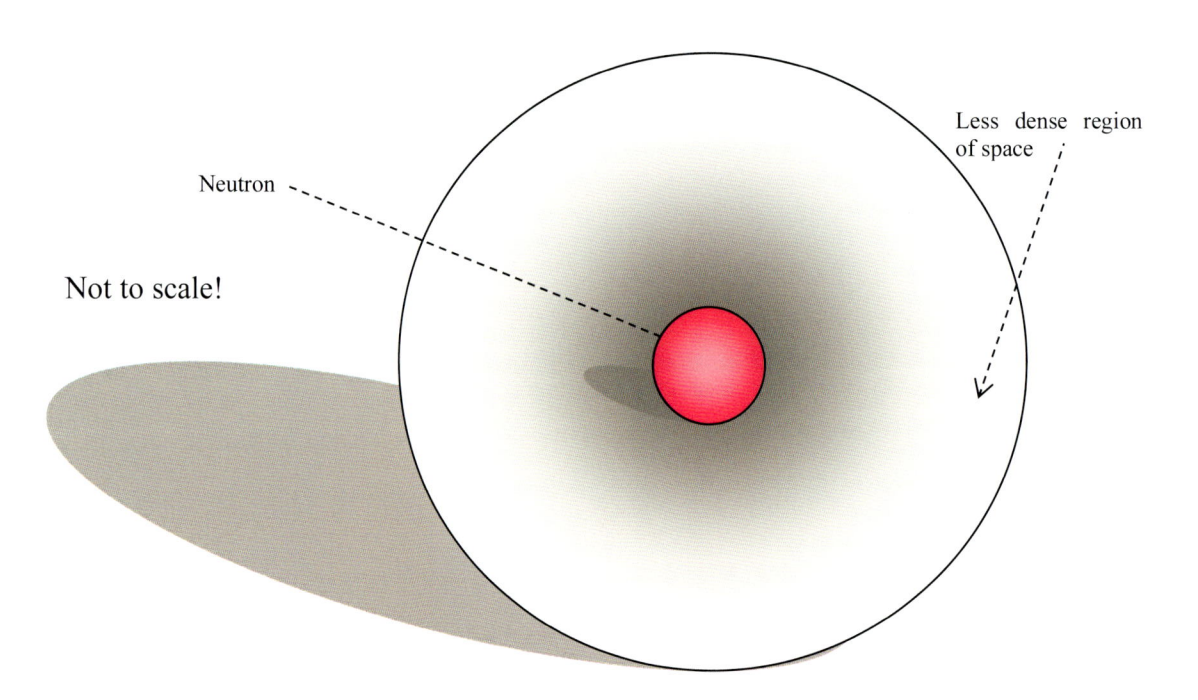

Also, the higher dimensional vortex is bent into a very tight loop. In this tight loop, the vortex is turned inside out, creating the weak force of nature [why the torus breaks – seen as the weak force - will be explained later]. Hence, the neutron's "neutral" charge, and the weak force of nature are both revealed, clearing up two more of the great mysteries of nature.

Because this extraordinary creation of nature we call the Neutron is so unique, I have attempted to illustrate it in the following drawings. Unfortunately, since it is impossible to draw 4d space, these 2d to 3d sketches are used:

Figure 10.4 INITIAL CONDITION:

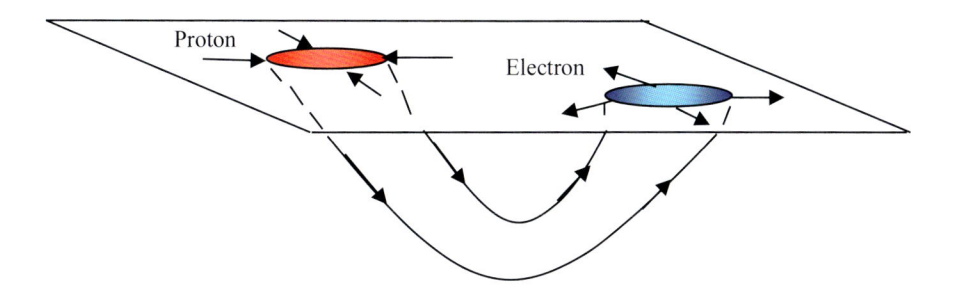

Figure 10.5 STEP #1: For any one of a number of possible reasons, the vortex between the proton and the electron breaks: [Note, this break is much closer to the surface than seen here.]

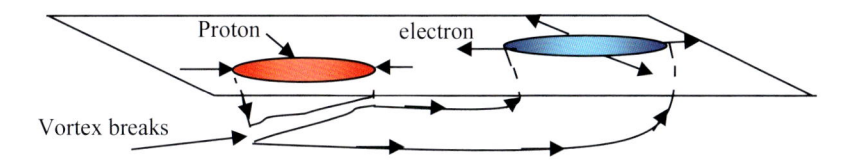

Figure 10.6 STEP #2: Isolating the proton from the above drawing, and expanding its size, note how the bottom of the vortex begins to curl outward:

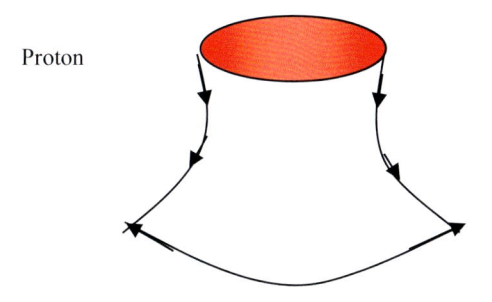

Figure 10.7 STEP #3: The curl becomes more pronounced as it continues to move upward:

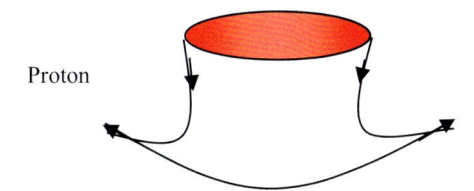

Figure 10.8 STEP #4: The vortex curls upward at its incredible velocity (the speed of light) towards the top of the hole we call the proton:

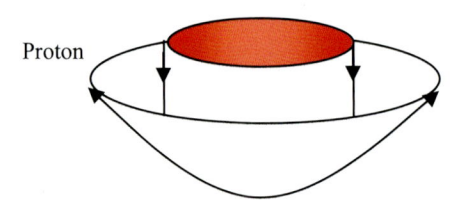

Figure 10-9 STEP #5: The vortex approaches the top of the hole called the proton:

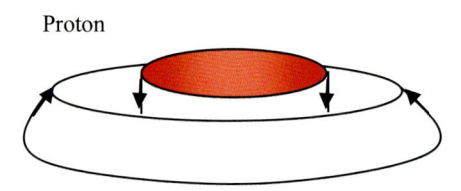

Figure 10.10 STEP #6: The vortex curls back into the hole called the Proton, forms a torus, the circulating flow begins and becomes a new "particle" science calls the Neutron. Note the 4d torus is impossible to draw so the following is a 3d torus.

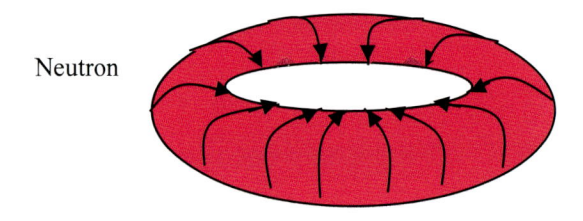

Figure 10.11 STEP #7: Another new "particle" called a *positron* is created when the end of the vortex that was attached to the electron reaches the two dimensional surface.

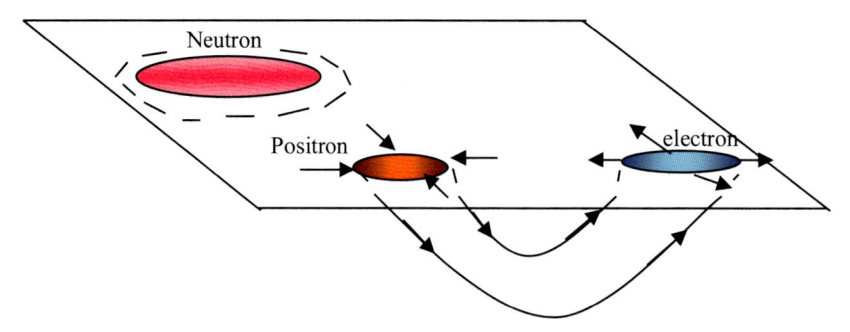

[Note how space now flows into the hole called the positron; turning it into a "particle" with a charge opposite to that of the electron. The free end of the vortex becoming a positron now explains the creation of the phenomenon called *Conservation of Charge* (from Books 2-3).

> The explanation of how the quarks change "Flavor" from the proton to the neutron will be explained in (Ch. 36).

Note: although the changes in the quark content from the proton to the neutron is explained later in the Ch.36, it does need to be mentioned here that the 0 charge on the neutron is not being created by its quark content. The quark content is created by the direction of space flowing into the proton via the twisted loop of the vortex going from the enveloping electron. How the charges are created on quarks and their relationship to the particles they inhabit was explained in Book 3.

The decay of the neutron will be explained when the weak force of nature is explained in Ch.15. [Just briefly, a discontinuity in the cross-section of the flow occurs, resulting in its break-up.]

This discontinuity is created at different times depending upon the density of the surrounding space. Lesser in smaller atoms with a lesser density of surrounding space; and taking much longer in more massive atoms possessing a greater region of less dense space.

> Just as a two dimensional surface (somewhat like a piece of paper) has two sides, our three dimensional space has two sides; the fourth, fifth, sixth, and seventh dimensional space all have two sides also. These two sides of each dimension create a number of extraordinary phenomena.

Chapter 11
The Two Sides of Space!

As we shall soon see, the number of "layers" of quarks reveals that there are at least seven dimensions in the universe. But just as important as the number of dimensions that exist are the number of sides of space that exist too.

In Books 1 & 2, the way to visualize the entrance into fourth dimensional space was introduced. It was noted that each higher dimension is at right angles to all of the lower dimensions simultaneously. For those who have not studied this relationship it can briefly be explained as follows:

Figure 11.1

The line represents a one dimensional line; the red square is a two dimensional plane at right angles to the one dimension line…

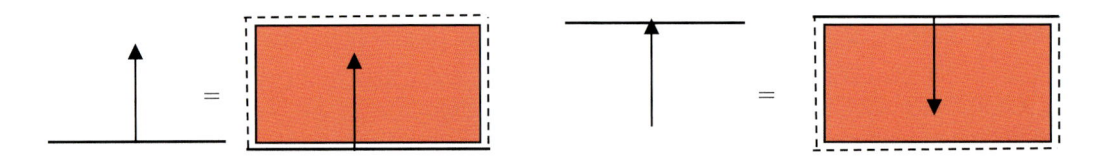

Figure 11.2

The third dimension is at right angles to the first and second dimensions simultaneously;

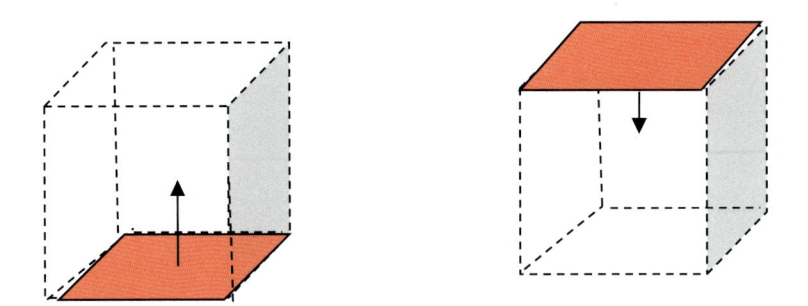

Figure 11.3

The fourth dimension is at right angles to the 3rd, 2nd, and the 1st simultaneously;

[This figure is impossible to draw. However, it can be represented by a series of arrows within a three dimensional sphere pointing towards or away from its geometric center.]

Figure 11.4 These figures are impossible to draw!

The fifth dimension is at right angles to the 4th, 3rd, 2nd, and the 1st simultaneously; the sixth is at right angles to the 5th, 4th, 3rd, 2nd, and the 1st simultaneously; and the seventh is at right angles to the 6th, 5th, 4th, 3rd, 2nd, and the 1st simultaneously.

[These figures are also impossible to draw.]

Looking at the above relationships, it is also easy to realize that each higher dimension possesses two different directions [one direction points towards the higher dimension, and the other points away from the higher dimension]. For the three lower dimensions these directions have already been defined: forwards and backwards, left and right, and up and down; and in Books 1 & 2, it was explained how the fourth dimension possesses two new directions that can be called "within", and "without". The importance of these opposite directions reveals that each higher dimension of space also possesses *TWO SIDES*!

Just as a two dimensional surface has two sides, our three dimensional space has two sides and the fourth, fifth, sixth, and seventh dimensional space all have two sides also.

The way to envision the two sides of any dimension is to imagine a hollow sphere. While viewing the sphere in your mind it is easy to see that there are two sides to its surface. One side faces the center of the sphere while the other side points away from the surface:

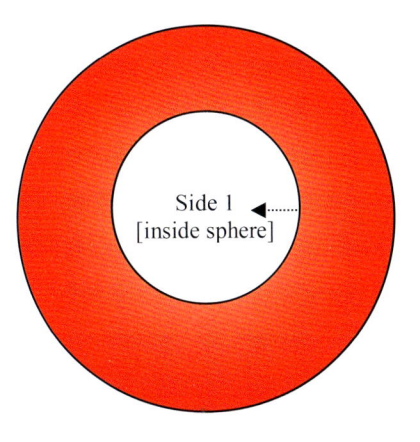

Figure 11.5   **Figure 11.6**

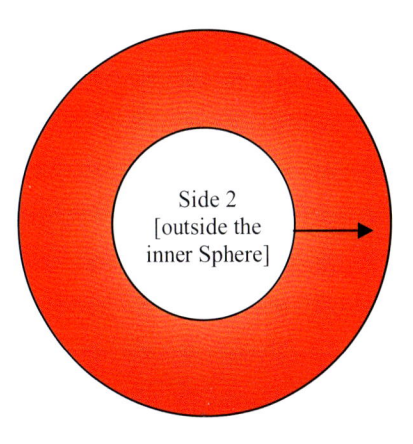

The side facing the center [the inside] can be called *Side 1*, while the opposite side [the outside] can be called *Side 2*. In the above figures, note how one arrow points to a three dimensional volume trapped within the sphere; while the other points to a three dimensional volume outside of the sphere. Using these two sides of space, we are almost ready to explain quarks; however, before we do, we must first introduce what can only be called, "the two volumes of space".

THE TWO VOLUMES OF SPACE

Many years ago, my preliminary investigations of the construction of quarks allowed me to hypothesize that we lived upon the expanding three dimensional surface of a volume of higher dimensional space. Until other evidence presented itself, because I did not know what this volume was expanding into, it was only logical to conclude that it was expanding into nothing at all - a void: but this was a mistake.

After the study of quarks began, I realized that the two different sized flowing volumes (charges) - the 1/3 and 2/3 charges - could be explained if two volumes of space were present instead of just one. These two volumes would be responsible for creating a mutually shared surface: just as the Earth's atmosphere and oceans share a mutual surface, a mutually shared surface would also exist between the two volumes of space. And equally important, if one volume was expanding into the other, then the surface of one volume would be in expansion and the surface of the other would be in contraction creating two different Elastic Moduli for each space: making one flowing volume (its charge), smaller than the other! Here is how it happens…

> Quarks are holes in higher dimensional space. Shockingly, they are holes within holes! Also, this relationship explains why quarks cannot exist outside of other "particles" such as the Proton: explaining the phenomenon called "Quark Confinement". [This idea that quarks are higher dimensional holes in space, reveals that the universe is constructed out of a number of dimensions; and gives us one of the first clues as to the actual construction of the universe.]

Chapter 12
Quarks Are Higher Dimensional Holes in Space

Just as protons and electrons are three dimensional holes existing upon the surface of fourth dimensional space, quarks are 4th, 5th, and 6th dimensional holes existing upon the surfaces of the higher dimensional space trapped within the interior of these three dimensional holes. To understand how this relationship works, it is extremely important to realize that each lower dimension is the surface of the next higher dimension.

Remembering the concepts introduced in Chapter 7 that reveal the relationships existing between a surface and its volume, suddenly, the following becomes easy.

Quarks are not solid particles, but higher dimensional holes in space! There are six quarks and three layers of them. The first layer are the Up and Down quarks. The Up and Down quarks are formed upon the surface of 4d space that exists within the 3d holes existing upon the surface of 3d space.

The following relationships between quarks and the space they inhabit reveals a "Hierarchy of Holes"! It begins with the three dimensional holes and progresses downward. In the following drawings, observe how the lower dimensional hole becomes larger as another higher dimensional hole is created within it.

THE HIERARCHY OF QUARKS

The Up quark is formed on Side 1, while the Down quark is formed on Side 2 (from Book 3). Both the Up quark and the Down quarks are 4d holes in the 5d volume of space. They exist within the 3d holes we mistakenly call electrons and positrons.

Figure 12.1   **Figure 12.2**

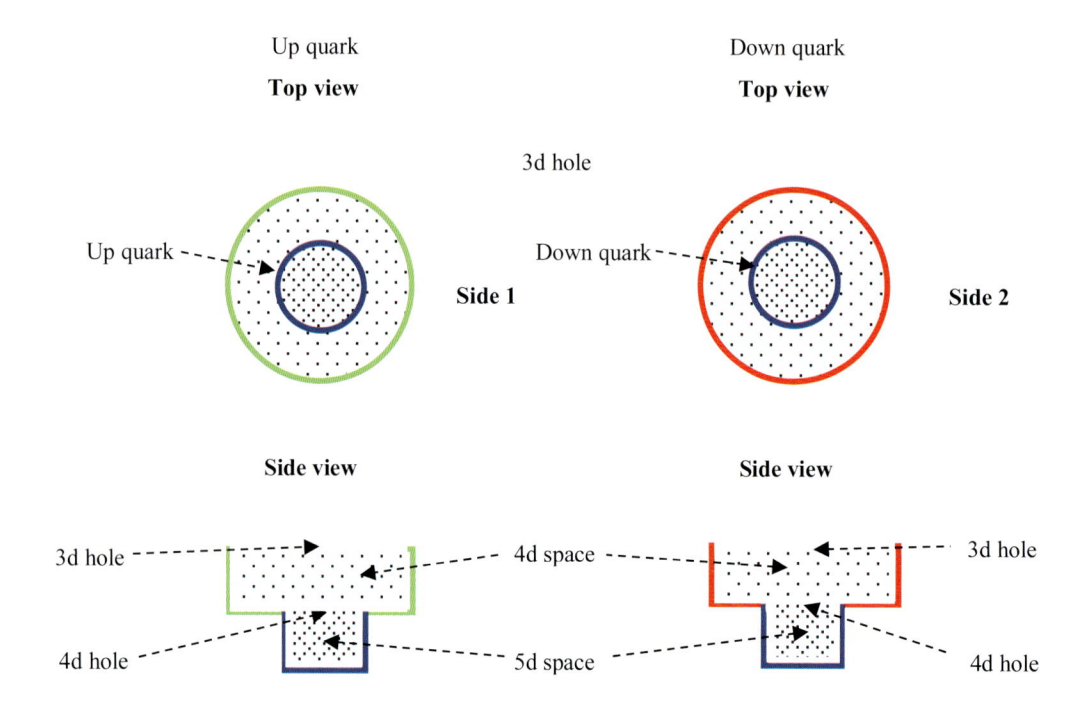

Up quark — **Top view**
3d hole
Up quark
Side 1

Down quark — **Top view**
Down quark
Side 2

Side view

3d hole
4d hole
4d space
5d space
3d hole
4d hole

The Charm quark is formed on Side 1, while the Strange quark is formed on Side 2. Both the Charm quark and the Strange quarks are 5d holes in 6d space.

Figure 12.3   **Figure 12.4**

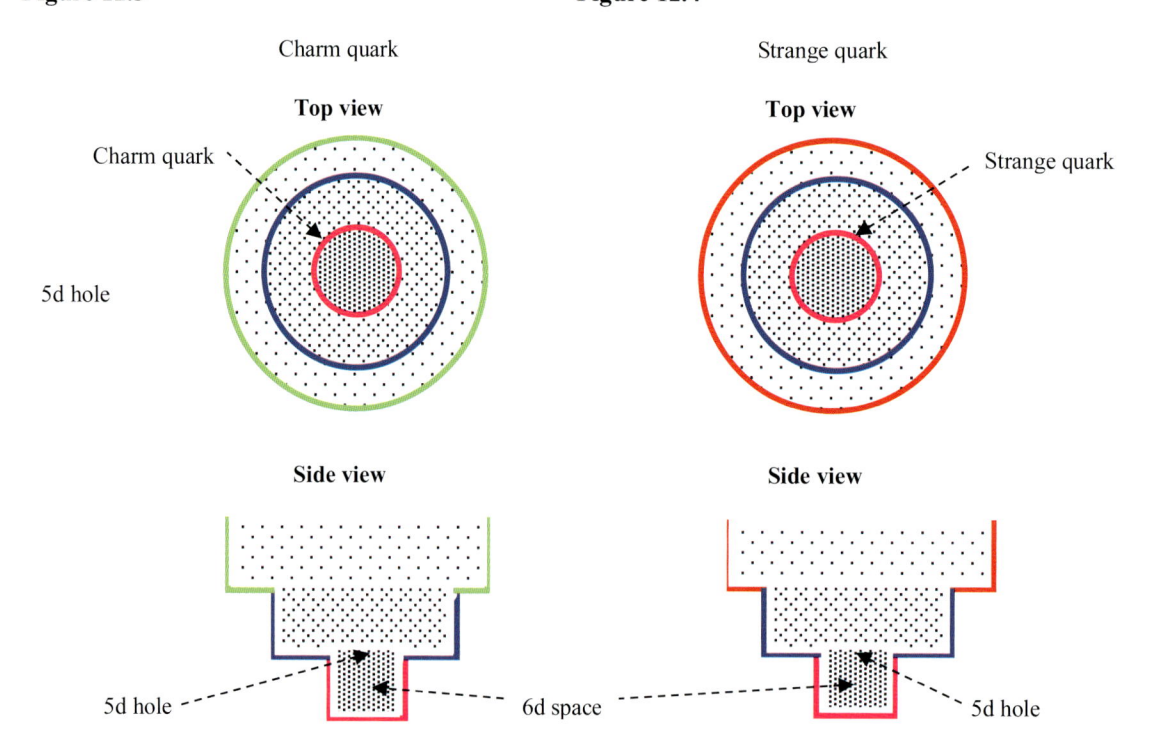

Charm quark — **Top view**
Charm quark
5d hole

Strange quark — **Top view**
Strange quark

Side view

5d hole
6d space
5d hole

The Top quark is formed on Side 1, while the Bottom quark is formed on Side 2. Both the Top quark and the Bottom quarks are 6d holes in 7d space.

Figure 12.5

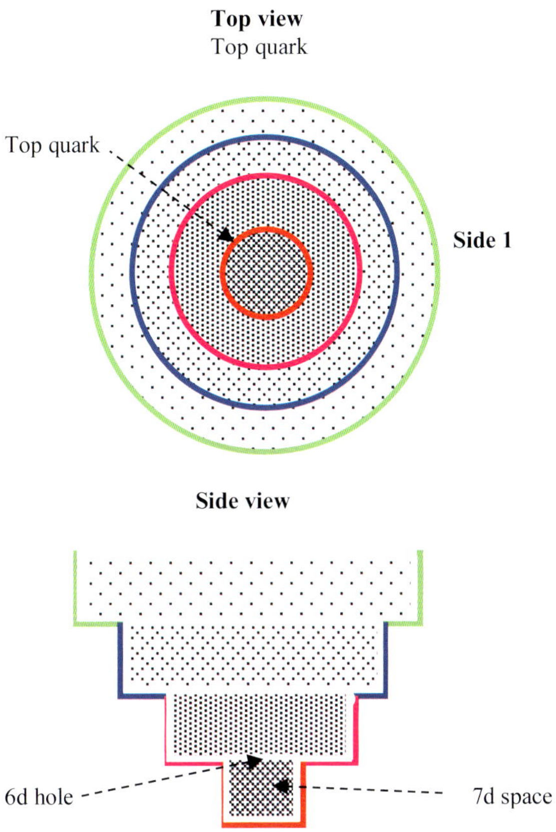

Top view
Top quark

Top quark

Side 1

Figure 12.6

Top view
Bottom quark

Bottom quark

Side 2

Side view

6d hole ---- 7d space

Side view

7d space ---- 6d hole

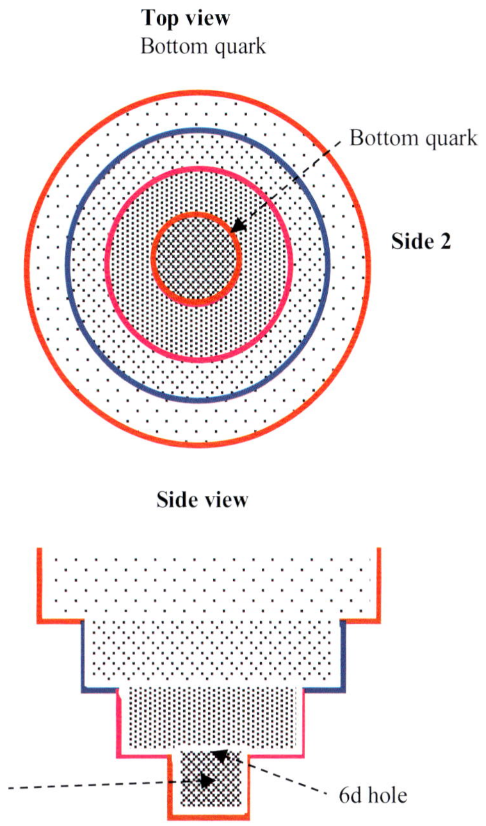

    Because a hole in a surface is an entrance into the next higher dimension, just as a two dimensional hole [such as a door, window, cave opening, pipe opening etc.] is the surface entrance into three dimensional space, a three dimensional hole is the three dimensional entrance into fourth dimensional space…

…a fourth dimensional hole is the entrance into fifth dimensional space…

…a fifth dimensional hole is the entrance into sixth dimensional space…

…a sixth dimensional hole is the entrance into seventh dimensional space…

…and so on depending upon how many dimensions exist. When higher dimensional holes exist within lower dimensional holes, this effect can be called "*Layering*".

    Collisions between particles open up holes in higher dimensional space. A fourth dimensional hole created upon the surface of fifth dimensional space is an *up* or *down* quark. If the collision is even greater, the warp can extend onto the fifth dimensional surface, tearing it and creating a hole upon the fifth dimensional surface extending into sixth dimensional space. This fifth dimensional hole creates another "layer" of quarks: the "*strange*" and "*charm*" quarks. And finally, if the warping is great enough, it can extend all the way into the sixth dimensional surface creating an entrance into seventh dimensional space. This final layer of holes creates the "*top*" and "*bottom*" quarks.

This idea is not speculation. It is based upon the decay of more massive particles into less massive particles (Book 3). The creation of higher dimensional holes "sheathed" within lower dimensional holes finally allows us to be able to explain the mystery of "quark confinement".

EXPLANATION OF QUARK CONFIMENTS:

Just as a fish is trapped within the water of a fishbowl, quarks are trapped within the volume of higher dimensional space sheathed within a lower dimensional hole. For example, what are called Up and Down quarks are fourth dimensional holes trapped within the fourth dimensional space sheathed within three dimensional holes; Strange and Charm quarks are fifth dimensional holes trapped within the fifth dimensional space sheathed within fourth dimensional holes; and Top and Bottom quarks are sixth dimensional holes trapped with the sixth dimensional space sheathed within fifth dimensional holes.

Using the above relationships, it is now clear why a quark cannot exist alone as a single "particle"- unique unto itself: a hole cannot exist outside its volume. Hence a fourth dimensional hole cannot exist outside of its fourth dimensional space.

Since our universe is really the three dimensional surface of a fourth dimensional volume of space, any hole penetrating into the fourth dimensional volume has to first begin upon the three dimensional surface. Because each higher dimensional hole has to be sheathed within its lower dimensional surface, as progressively higher dimensional holes are created, they can only be created after a previous hole was first created upon their lower dimensional surface. This creates a hierarchy of holes forcing each higher dimensional hole to exist within a lower dimensional one.

> The particles of nature exist unlike anything 20th Century science realized or proposed.

Chapter 13
The "Particles" of Nature Containing Quarks Are Briefly Explained

I wanted to explain the forces of nature at this point in this book. But to do so entails the explanation of "particles" containing quarks. So here goes. This is but a brief explanation that will allow us to explain the forces of nature.

What first must be understood is that these so-called "particles" are not particles at all. They are holes in space!

THE ELECTRON

Everything begins with the electron and the positron. The electron is a three dimensional hole in space connected to its anti-particle the positron by a fourth dimensional vortex. In reality, these two holes are merely the ends of the vortex of flowing space, much like the ends of a pipe.

Figure 13.1

The electron and the positron are connected by an invisible vortex of 3d space flowing from the positron to the electron in 4d space.

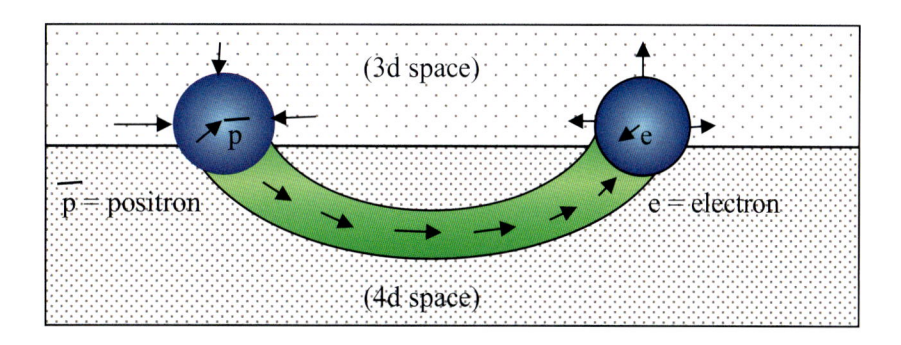

Figure 13.2

The quarks are the higher dimensional ends of invisible vortices of higher dimensional space flowing between them. Below, is an Up quark Anti-Up quark combination. Notice how the Anti-Up quark is identified by a little line above it.

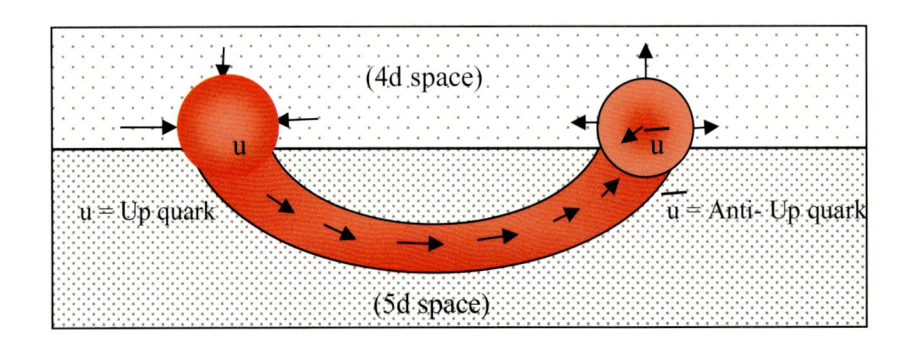

MESONS

Mesons are "particles" that contain two quarks. These quarks are "sheathed" within the three dimensional holes called the electrons and positrons. There are many types but the most common are the Pions. Pions with Up and Down quarks; typically possess 1/7 the mass of the proton. Just as protons and electrons have anti-particles, so do quarks. Contrary to present belief, the quarks do not create the electrostatic charges of the particles they inhabit.

The + 1 or – 1 charges are created by the positron or electron they are sheathed within. This will be explained when charges are explained.

Figure 13.3 **Figure 13.4**

The Pion   The Anti-pion

Sheathed within an enlarged positron

Sheathed within an enlarged electron

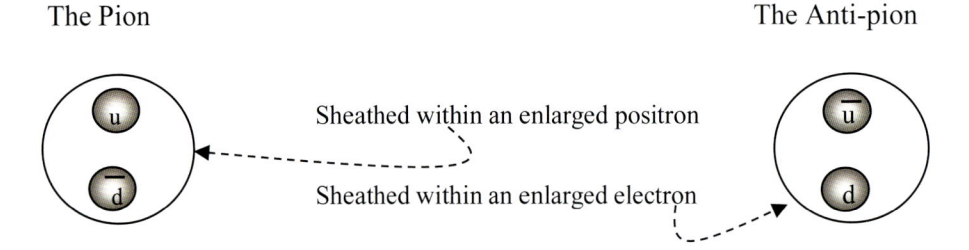

BARYONS

Baryons are "particles" that contain three quarks. The most well-known are protons and neutrons.

PROTONS

Space flows into the proton and out of the anti-proton. Just like the mesons, the three quarks of each of the below particles are sheathed within positrons and electrons. Hence, their charges create the +1 or – 1 charges of these two particles and not the quarks inside of them.

Figure 13.5   **Figure 13.6**

Proton   Anti-Proton

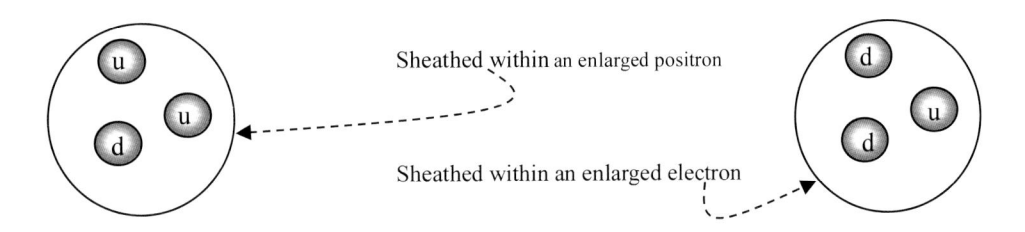

Sheathed within an enlarged positron

Sheathed within an enlarged electron

NEUTRONS

As explained previously, neutrons are constructed out of protons and electrons. They have no electro-static charges because the 4d vortices are bent into sharp loops. And again, the quarks do not create their 0 charge. This will be explained later when charges are explained.

Figure 13.7   **Figure 13.8**

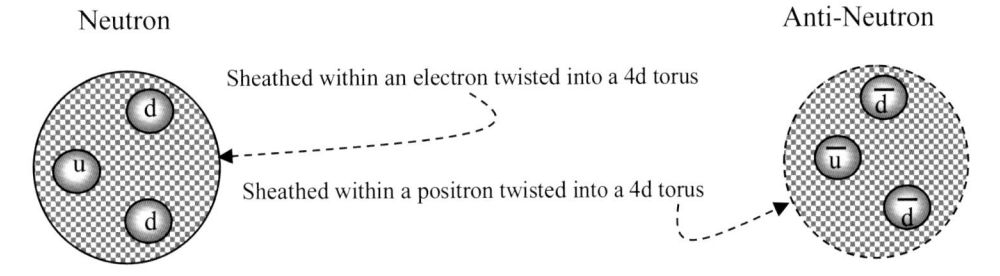

Neutron   Anti-Neutron

Sheathed within an electron twisted into a 4d torus

Sheathed within a positron twisted into a 4d torus

Now we are ready to explain the forces of nature. This would be a dry mundane explanation if it was not for the spectacular discovery regarding the Strong force in nature!!!

PART IV

THE FIVE FORCES OF NATURE

> The Five forces of nature are explained via the Vortex Theory of Atomic Particles: the Strong force, the Weak force, the Electromagnetic force, the force of Gravity, and the newly discovered fifth force the Anti-gravity force!

Chapter 14
The Strong Force…

Unlike the 20th Century explanation for the Strong force in nature that is revealed to be a mistake; the true explanation for this powerful force is a fantastic, spectacular discovery! Rivaling anything ever discovered before. [This is why I love science!]

The true explanation for the Strong force was "preliminarily" discovered in 1935 by the great Japanese physicist Hideki Yukawa with his mathematical proof of a Virtual Particle being passed back and forth between particles in the Nuclei of atoms. Unfortunately, he was decades ahead of his time. His was a great idea → a brilliant revolutionary idea! But unfortunately he had no idea how it worked. His discovery would have to wait for the discovery of quarks, mesons, and the discovery of the Vortex Theory to explain how the exchange of quarks between particles are creating this powerful force.

PROBLEM WITH THE CURRENT VISION OF THE STRONG FORCE

The current vision of the universe proposes that particles containing quarks are held together by other proposed particles called gluons. It is further proposed that this force increases with distance! An idea that is contrary to everything we have ever learned about the strength of fields dissipating in strength with the square of the distance from their source.

This outrageous proposal was offered to explain the phenomenon of Quark Confinement; and to explain the observation that no quark has ever been seen outside from the particle it is contained within. We know now that all of this is a mistake. The Vortex Theory of Atomic Particles reveals that quarks are higher dimensional holes existing within three dimensional holes and cannot leave the surface of the higher dimensional space they exist within.

DO GLUONS EXIST?

Quarks possess higher dimensional charges that can flow between higher dimensional holes, and can be called gluons for lack of a better word; but gluons cannot explain what is holding together the larger 3d holes called the protons and neutrons they exist within.

Hideki Yukawa's virtual particle

In 1935, Hideki Yukawa proposed that the strong force could be explained by the existence of a virtual particle approximately 200 times the mass of the electron [1/7 the mass of the proton] being passed back and forth between protons and neutrons in the nucleus of the atom.

This particle was designated a "virtual particle" because it passes back and forth so quickly it cannot be seen or detected. At the time of its proposal, this idea was highly speculative because no particle of this mass was known to exist. But in 1947, a team of physicists lead by Cecil Powell from Bristol University in England discovered the Pion. The Pion [a particle we now know contains two quarks: usually an Up and a Down] appeared to be exactly the particle predicted by Yukawa, and it was 1/7 the mass of the proton! However, even though such a particle was discovered to exist in nature, the problem of that era was that quarks had not yet been discovered! So nobody knew how it could exist within protons and neutrons; or how it could be passed back and forth between protons and neutrons; and how come it moves so fast it cannot be seen?

THE EXPLANATION OF YUKAWA'S VIRTUAL PARTICLES

According to the Vortex Theory, when a neutron first approaches a proton, the less dense space surrounding each "particle" distorts the shape of the other. Then, as both distorted holes attempt to straighten out, they move slightly in the direction of the other. As this process continues, each additional movement increases the velocity of the particles, accelerating each in the direction of the other until they collide. As they collide, these two holes try to bend into each other, and as they do, some of the space flowing around and around in the neutron's fourth dimensional [4d] torus tries to flow into the fourth dimensional [4d] vortex of the proton.

This movement of space out of the torus causes it to break. As the torus breaks two ends are created: one end connects to the proton's 4d vortex; the other end emerges back into 3d space, turning it into a proton. At the same instant, the broken end of the original proton's 4d vortex now wraps around and envelops *it*, changing its vortex into a torus; changing its identity - "metamorphosing" the proton into a neutron; (and the neutron into a proton). However, just as soon as the switch occurs, the process instantly begins all over again, causing the newly formed "particles" to revert to their former identities. This constant switching of identities keeps one "particle" pressed tightly against the other – becoming the strong force of nature.

Figure 14.1

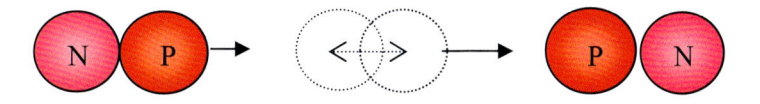

THE CREATION OF YUKAWA'S VIRTUAL PARTICLES

Even though the creation of these continual and seemingly instantaneous microscopic metamorphoses is a fascinating and dramatic event, an even more remarkable event is taking place within the proton and neutron. As the proton and the neutron constantly merge together, change identities, and split apart, two quarks are continually being passed back and forth between them.

As the two particles merge together, an up quark from the proton is passed to the neutron, and a down quark from the neutron is passed to the proton. And as this transference occurs, as the two different quarks pass by each other on their continuous journey to the opposite "particle", for a brief instant as they pass each other, a virtual pion is created! [How fascinating!]

In Figure 14.2 below, the proton and neutron are touching; in Figure 14.3, as the proton and neutron begin to change identities, an up quark from the proton and a down quark from the neutron approach each other; in Figure 14.4, as the proton and the neutron merge together, the up and down quarks pass each other creating for a brief instant a **Virtual Pion**; and finally, in Figure 14.5, after the transformation is complete, the neutron and proton have switched identities.

Figure 14.2

Figure 14.3

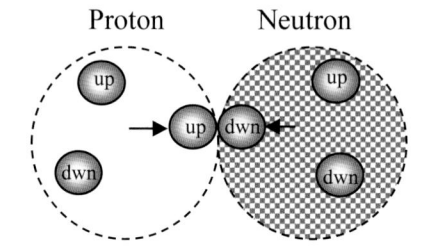

Figure 14.4

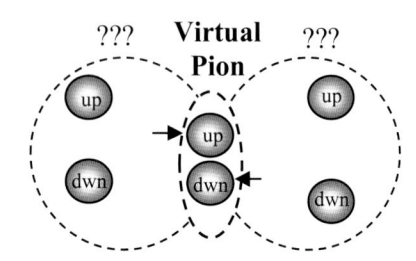

Figure 14.5

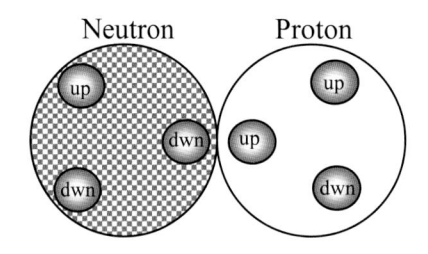

Consequently, Hideki Yukawa's proposal of a virtual particle being passed back and forth between the proton and the neutron – before the discovery of quarks and pions, is simply a stupendous achievement; and he needs to be posthumously recognized for this great discovery!!!

The Strong Force...Part II

The first half of the strong force holds protons and neutrons together. Alpha particles are also held together by the constant switching of identities of all four particles [Ch 21].

However, there is a second half of the strong force that contemporary science knows nothing about. This second half of the strong force is created by quantum gravity. Quantum gravity is what holds the alpha particles, the proton neutron pairs, and the neutrons together in the nuclei of atoms.

It is fascinating to report that the combinations of the alpha particle's close proximity to each other causes their nuclear gravity to create gravity "wells" of less dense space that new proton neutron pairs and new alpha particles are drawn to, and "fall into" causing all atoms in the universe to be constructed exactly like all other similar atoms in the universe! Creating an order and harmony, that is unique unto itself! What a beautiful universe!

Because this takes a book all its own to explain how the addition of new alpha particles to existing nuclei creates these wells and builds up the atomic structures of nuclei, it will be explained in a seventh book devoted totally to the construction of the elements of nature.

This work will also explain how the anti-gravity effects of the denser spaced vortices flowing back from the electrons and into the interior of the nuclei of atoms tends to make these heavier nuclei unstable and is responsible for keeping nuclei from obtaining gigantic sizes.

Like the strong force, the Weak force is also created in a manner conventional science is unaware of. The W particle and the Z particle that supposedly "moderate" the weak force are also mistakes! The truth about what is really occurring is most fascinating!

Chapter 15
The Weak Force...

The weak force of nature is not really a force. The weak force of nature is associated with the neutron and how it "decays" into a proton, an electron, and an anti-neutrino. However, the "decay" of the neutron is not caused by a "force", but rather by the break-up of the three dimensional vortex surrounding the proton located at the center of the neutron! This break-up is caused by harmonics within the tightly bent vortex as it whirls in and out of the proton at the speed of light.

THE TIGHTLY BENT LOOP OF THE VORTEX...

The weak "force" is a result of the higher dimensional vortex being inverted into a tight loop. Because this loop is in the fourth dimension it is impossible to draw, therefore the following is only a representation.

Figure 15.1

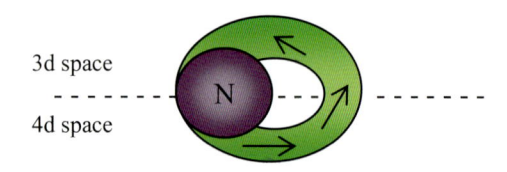

This tight loop is seen better in Figure 15.2 below as 3dspace flows from the proton back to the electron in 4d space.

Figure 15.2   As seen from outside the Neutron.   **Figure 15.3** Cross-section of the Neutron

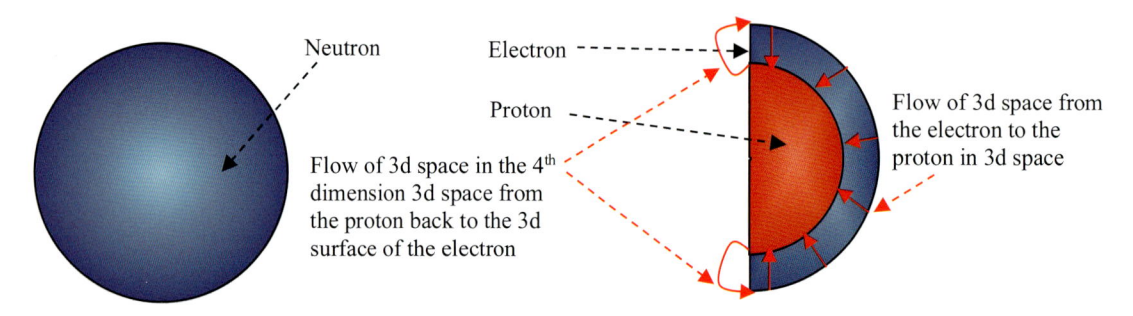

And just like the water flowing around a bend in the river, the outer flow is moving faster than the inner flow. In over hundreds of trillions of rotations taking place in just 10.5 minutes, the vortex becomes lopsided with more space flowing on one side than the other. This unbalanced situation causes the vortex to break…

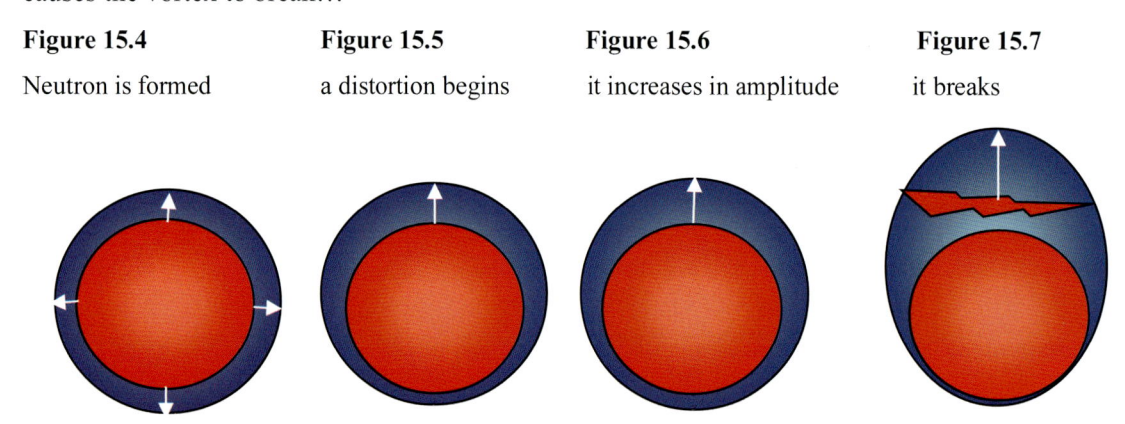

Figure 15.4	**Figure 15.5**	**Figure 15.6**	**Figure 15.7**
Neutron is formed	a distortion begins	it increases in amplitude	it breaks

When this tight loop of flowing three dimensional space is broken, the ends of the vortex are again separated, allowing the proton and the electron to seemingly, magically reappear.

What causes the vortex to break is the denser space surrounding the electron trying to move outward and away from the space bent inward surrounding the proton. Because the tight loop is an unnatural bend, the space within the vortex is stretched more on the outside than it is on the inside of the loop. This stretched condition makes it less elastic, decreasing its elasticity.

Within the vortex, this decrease in elasticity makes its inside bend want to flow at a faster rate than its outside bend. This creates a stress between the inside edge of the flow and the outside edge. As the inside of the vortex tries to flow faster, it tries to pull away from the outside of the vortex, and eventually, the construction of the vortex is unable to handle the strain, and it breaks along its outside edge...

Figure 15.8   **Figure 15.9**

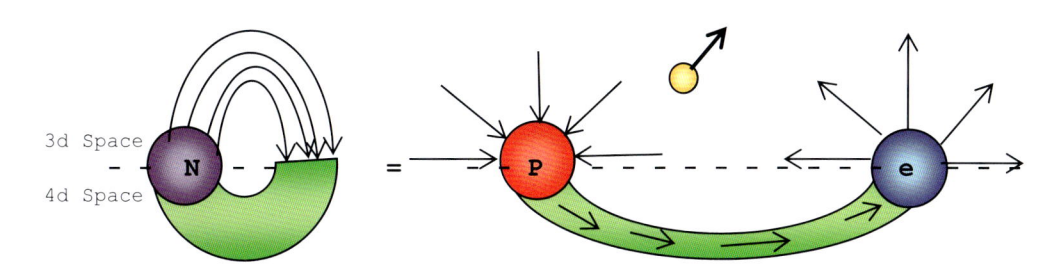

In Figure 15.7, note how the vortex breaks at the neutron; then in Figure 15.8, note how the proton reemerges at one end of the vortex while the other end of the vortex becomes the electron; as an invisible electron-antineutrino flies off [the neutron is the extra fourth dimensional space needed to increase the volume of the electron enabling it to encircle the proton].

When the vortex breaks, the two ends of the vortex reappear upon the three dimensional surface as the proton and the electron. The anti-neutrino "particle" is created by the deflation of the volume of the three dimensional space that the larger neutron filled. The whole process can be compared to twisting a strong spring into a sharp bend, letting go of it, and watching it snap back into its original shape.

Even though the weak force is more of a disturbance than a force, it is interesting to note that both the electromagnetic force and the weak force are related because they are both created out of flowing space.

Also, when the neutron is alone in free space, it only lasts about 10.3 minutes before it "decays" into a proton, an electron, and an anti-neutrino. But when the neutron is within the nucleus of an atom, it lasts much longer. The reason why it survives so much longer comes from the fact that space near the nucleus is less dense, making the vortex flow slower, making it take longer for the harmonics to develop.

THE MISUNDERSTANDING ABOUT WHAT THE W & Z PARTICLES ARE

The present day belief that the W and Z particles transfer the weak force is totally and completely wrong. Nothing could be further from the truth. The W and Z "particles" are the breaking vortices. In Figure 15.7 above, note how the vortex breaks at the neutron. Before the electron has a chance to form, the breaking vortex is what science has mistakenly called a "particle." Its positive or negative charge depends upon if it is a neutrino or anti-neutrino whose vortex breaks. Both the W particles and the Z particles are explained in Chapter 32.

> The electromagnetic force is pretty much like it is described by contemporary science. There are no revolutions here except to say it is created by flowing space. The force is split up into two parts: as an electrostatic force; and as an electromagnetic force.

Chapter 16
The Electromagnetic Force...

THE ELECTROSTATIC FORCE...

The electrostatic force is created by space flowing into or out of three dimensional holes. The electrostatic force is created by three dimensional space flowing into the proton and out of the electron;

Figure 16.1

Electrostatic Force

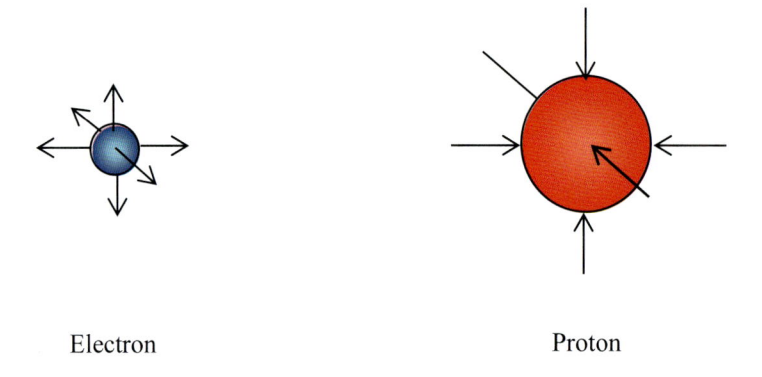

Electron   Proton

THE ELECTROSTATIC "ATTRACTIVE" FORCE...

The electrostatic "attractive" force develops when an electron and a proton come into a close enough proximity to each other that flowing space from one hole begins to flow directly into the other hole. The mutual motion towards each other occurs because the space flows more readily out of the electron in the direction of the proton – distorting the electron in the direction of the proton – causing it to move in the direction of the proton; while the flow into the proton occurs more easily in the direction of the electron – distorting the proton in the direction of the electron – causing it to move in the direction of the electron. It is the distortion of each "particle" [seen in Figure 16.2] that moves them towards each other, and NOT the attraction caused by some mysterious force!

Figure 16.2

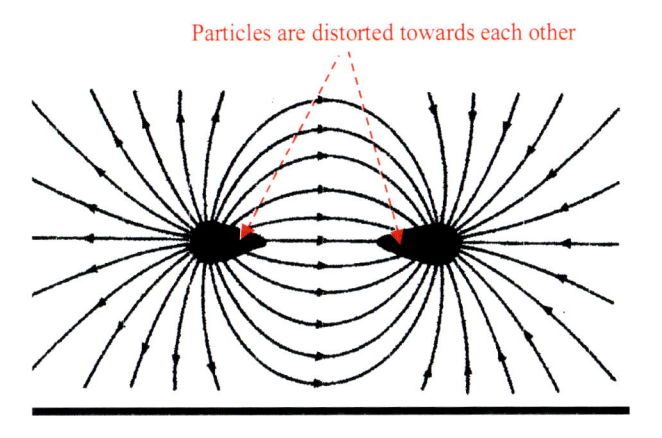

Particles are distorted towards each other

THE ELECTROSTATIC REPULSIVE FORCE BETWEEN TWO ELECTRONS…

The electrostatic repulsive force develops when two like holes such as two electrons come into close enough proximity to each other causing the flow of space directly between each hole to be "twisted" outward. For example…

When two electrons come into close proximity to each other the outward flow of space from one hole pushes against the space flowing out of the other hole. This resistance makes the region between each hole appear to be a denser region of space. This seemingly denser region causes the side of each hole directly opposite to each other to bend outward easier than the side facing it. This creates a pear shaped distortion in the surface of each hole. These distortions cause each hole to try to straighten back out into a sphere: in effect, causing each hole to accelerate in the opposite direction to each other; forcing them to move away from each other. Note: the greater the charge, the greater the distortion, and the greater the acceleration.

Figure 16.3

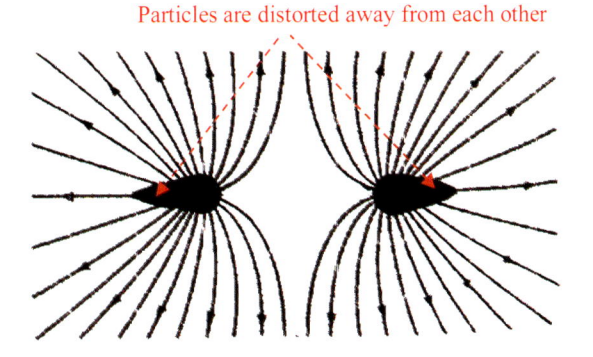

Particles are distorted away from each other

THE ELECTROSTATIC REPULSIVE FORCE BETWEEN TWO PROTONS…

Although it is easy to see why two electrons "push" themselves away from each other, why two protons repulse each other is not readily apparent.

When two protons come close to each other, it first appears as if the space flowing into each will pull them into each other. However, the reason why they don't; comes from the fact that the three dimensional space between them cannot flow in two opposite directions simultaneously.

Consequently, to be able to flow into both particles along the x- axis, space has to now flow downward from the direction of the +y axis, and upward from the direction of the –y axis to get into each hole. This situation creates the exact same effect as when space was flowing out of both electrons creating a denser region of space between them. Because space does not flow into one side as easily as the other, the space directly between the two protons now creates the effect of a denser region of space. Hence, the opposite sides of each hole bend out easier than the sides facing each hole. They distort outward easier, and this outward bend accelerates them away from each other.

Figure 16.4

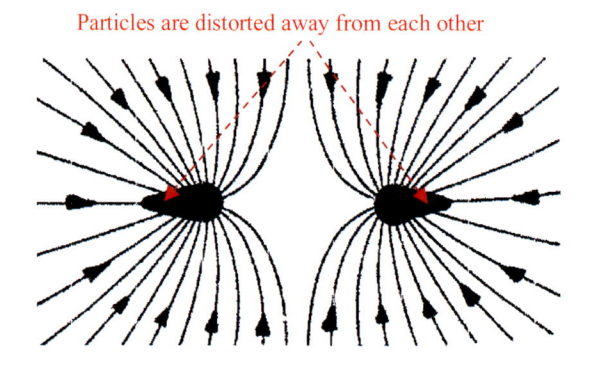

Particles are distorted away from each other

THE MAGNETIC FORCE…

Fortunately, for the magnetic force, there is nothing revolutionary to propose. It is fairly easy to understand, that Magnetic fields are created by rotations of three dimensional space around electrons. The spin of electrons creates rotations in space about them creating magnetic fields. It should be noted that the "attraction" and "repulsion" of magnets create the same distortions within protons and electrons as those seen in Figures 16.2 to 16.4 above. The additions of their spins in bar magnets create the famous lines of flux seen below in Figure 16.5.

Figure 16.5

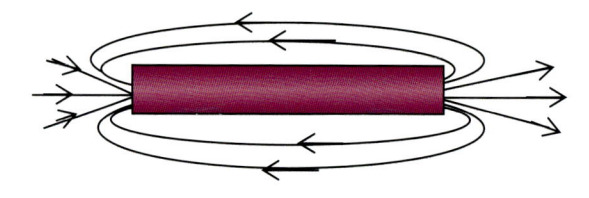

Figure 16.6

Top view   CW   CCW

The two spin states of electrons shown above are either clock-wise or counter-clockwise.

THE ELECTROMAGNETIC FIELD ABOUT A CONDUCTOR...

When current flows in a wire, free electrons in the wire move from one atom to the next. As the electron moves, a point is reached where it breaks contact with one vortex while at the same instant begins to make contact with a vortex from the next atom along its line of travel.

During this brief instant, some of the space flowing outward from the electron is allowed to stretch out to other atoms lying transverse to its line of travel. It is the combination of all of these momentary movements that is responsible for the electro-magnetic field about a conductor. The direction of the flows (left hand rule) is determined by the spin of the electrons. The spin of all the electrons is the same because their magnetic moments are aligned to the direction of the electric potential connected to the wire. The strength of the flow – the amount of flowing space – (magnetic field) depends upon the amount of electrons in transit at any one particular instant: which is a function of the amount of current in the wire (Coulombs per second).

So why don't particles move like logs in rivers if the electromagnetic force is flowing space?

A NEAT TRICK OF NATURE! WHY MATTER DOES NOT FLOW IN THE MINIATURE RIVERS OF FLOWING SPACE!

We all know that when we throw a stick in the river, it flows along with the river like it is a part of it. So, if electrostatic and magnetic charges are really flowing space, how come when matter is introduced into the flow, it does not flow too, like the stick?

It doesn't because of a neat trick of nature!

Long ago, when I first started upon this quest, and made the discovery that electrostatic and magnetic lines of force were really tiny flowing "rivers of space", I was bothered by the fact that if so, then why didn't matter flow along with these miniature rivers?

It took many years before I found the answer, then it was a conundrum, but now it all seems all too easy. There is the explanation...

The key to understanding what is happening is to remember that all matter: protons, electrons, and neutrons are holes in space. Consequently, when say a proton is part of the nucleus of an atom, [say a hydrogen atom for example], its electric charge is neutralized by the electron. Hence it can now be just considered to be a hole in space: a hole in space that space is constantly reconfiguring around it as it moves.

So when a powerful electrostatic charge is brought close to it, a strange and wonderful effect begins to occur. The flow of space in the direction of the atom wants to get around the hole. In the figure below, the circle below represents a cross section of a proton, while the arrows represent the electrostatic field:

Figure 16.7

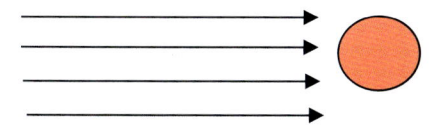

At first glance it appears as if the flowing space will push the proton off to the right of the page. However, this doesn't happen because the hole is a void in space. So instead of pushing the hole, the flowing space has to flow around it:

Figure 16.8

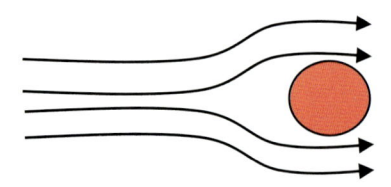

However, when it flows around the proton, it has to split.

Figure 16.9

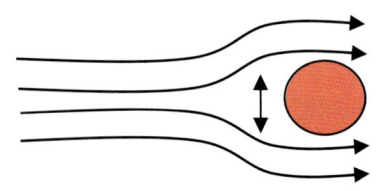

And when the space splits apart, the hole that is the proton is distorted in the direction of the split:

Figure 16.10

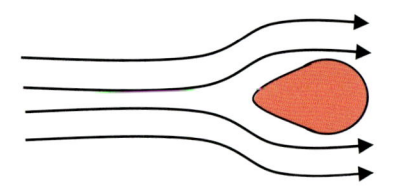

This distortion accelerates the proton in the direction of the flowing space. The greater the volume of the flow [the greater the flux], the greater the distortion: and hence the greater the acceleration. Maintaining its position, keeping the proton from being swept away! A simply neat trick of nature!

This same situation happens with magnetic fields. Also, a little thought will make one realize how the opposite happens when the field is reversed. This discovery will make one feel good: giving one the confidence born of success! A victory upon the playing field of science!

> The force of gravity is one of the most important of all the discoveries made using the Vortex Theory of Atomic Particles because it clears up one of the great misconceptions of how the force of gravity is really formed: it is not a Pull, but a Push!!!!!!!!!!!!!!!!!!!!!!!!!!!!!!!

Chapter 17
The Force of Gravity

In the theory of Relativity, Albert Einstein believed that matter was made of something, and space was made of nothing. And yet, he also believed that gravity was created by bent space surrounding stars and planets. *But how can something made of nothing be bent?*

Fortunately, the truth is now known. Although space appears to be bent around planets and stars, this is an *effect* being created by a region of less dense space and is not the cause of gravity; it is not bent space that is creating the force of gravity. The force of gravity is created by a massive volume of less dense space!

Most scientists and engineers are told in college courses that the force of gravity is created by a particle called the graviton. But nothing could be further from the truth! If it was, the amount of gravitons having to exist between all of the molecules of all the matter in the universe would exceed and overwhelm all the matter in the universe. Also, if so many of them did exist, they would literally fill space. So how come NO graviton has ever been seen in all of the trillions of collisions in all of the Linear Accelerators ever built in the world!!!

The answer is easy, they don't exist. It is a failed proposal of yet another failed proposal → Particle Logic!

It is the less dense regions of space that surround the holes ("particles") of matter that create the force of Gravity. Einstein's hypothesis that bent space is equivalent to gravity is wrong. The "bent space" surrounding stars is an *effect* created by the addition of all the spherical shells of less dense space surrounding all the protons and neutrons that account for the majority of the mass in stars and other large astronomical bodies.

When we stand upon this planet, this less dense region of space surrounding the Earth distorts the shape of every proton, electron, and neutron within our bodies towards the Earth's center of mass. The collective attempt of these particles to straighten out - pushes us towards the surface of the Earth and becomes the "force" we identify as our weight. Shockingly, *we are not attracted towards the earth*; the particles that atoms are made out of **PUSH** themselves towards the center of mass of the earth!

Figure 17.1

The Earth's Gravitational field: Although the big sphere represents the Earth's gravitational field, this is only a representation. Because the field is massive in relation to the size of the Earth, this smaller field was drawn to fit the page.

Note the distorted spherical brown hole above the Earth [out of proportion], this hole represents a single proton in the body of a person. Although the proton is a three dimensional sphere, it is now distorted into a pear shape; causing it and the rest of the protons and neutrons the person is made out of to be accelerated towards the center of mass of the earth.

Because the sphere surrounding the Earth represents the three-dimension volume of less dense space that is stretched inwards towards the center of the Earth, the proton is now caught in this less dense region too. But there is now a problem.

As the three dimensional space surrounding the Earth is stretched towards the Earth, it cannot stretch across the fourth dimensional void within the center of the proton. Consequently, the side of the proton facing the Earth is stretched towards the Earth while the side facing away from the Earth still retains its spherical shape:

Figure 17.2

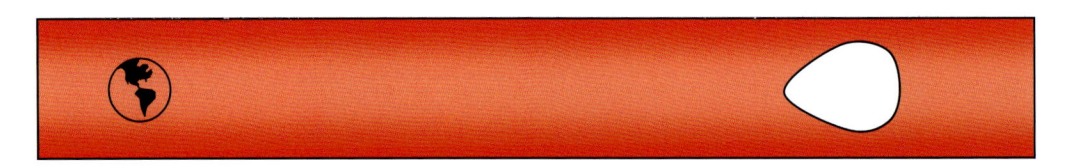

Although the above picture is a not a Picasso, you get the idea. You can see how the shape of the proton is distorted in the direction of the Earth. This distortion of the three dimensional surface of the three dimensional hole is responsible for creating the force of gravity. It is now understood that gravity is not a "Pull"; instead it can now be seen that surface distortions of the "particles" that make up matter "Push" them towards other matter. As can be seen in the diagram below, the distortion of protons in the nucleus of two different atoms are **not** "attracted" to each other, but are accelerated towards each other by their mutual distortions:

Figure 17.3

(Note: these two holes represent two protons in the nuclei of two different atoms.)

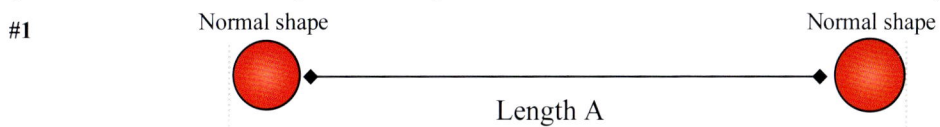

#1   Normal shape   Normal shape

Length A

When two protons are near enough to each other, the less dense regions of space that surround them (these regions are not shown here) overlap, creating an *even less dense region of space directly between them.* This less dense region of space surrounding each sphere stretches the surfaces of the two spheres that are directly opposite each other, but cannot stretch upon the opposite sides of the spheres. This effect distorts the sides of the two spheres that face each other into "pear shapes":

#2

At the very tip or point of the pear shape, space is sharply bent. This stress creates a strain upon the front surface of the hole, forcing the back of the hole to move forward: causing it to move forward.

#3

As the back of the distorted hole moves forward, the spherical shape of the hole is again created: and the proton is now moving at velocity v_1.

#4

However, the instant the hole begins to move forward, regaining its shape, the hole is distorted again; surface stresses at a and b cause such severe distortion that the surface of the hole at point c moves forward again to reduce the strain.

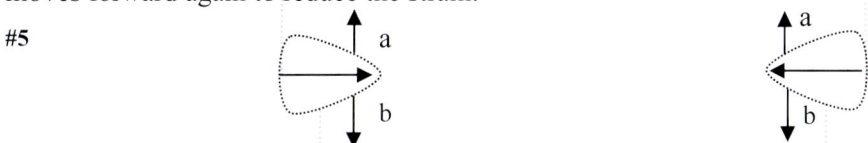

#5

The moving hole returns to the shape of a sphere, and the whole process begins again. But this time, this additional velocity added to the initial velocity now creates an acceleration: $v_1 + v_2$ = acceleration. And as it continues to accelerate, the velocity gets larger and larger: $\Sigma = v_1 + v_2 + v_3 + v_4 + n$. Where n = number of distortions.

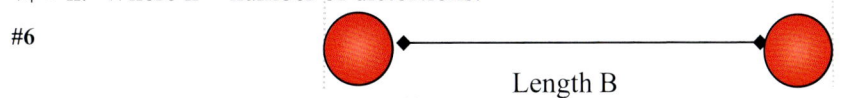

#6

Length B

Length B is now shorter than length A. Equally important, because the distortion of the three dimensional hole distorts the fourth dimensional space within, and subsequently distorts the surfaces of the quarks, the same acceleration is created upon the quarks too. This subject is discussed in Book 3, *The Quark Theory*.

As the process of distortion and reshaping repeats itself, length B becomes shorter than length A moving the two holes towards each other. As the distance between them shortens, the density of the space between them becomes even less dense. The decrease in density allows the hole in (#5) to elongate further, allowing the back to move upwards farther (#6) covering more distance, "accelerating" the holes towards each other.

The motions within the two protons disclose one of the great secrets of nature: the creation of internal waves within and surrounding subatomic "holes": see Chapter 27, Figure 27.7.

Because matter is made of holes instead of particles, it is extremely important to understand that a hole possesses an internal surface while a particle does not. *Hence, a hole can have a wave created upon its internal surface* while the particle cannot.

The realization that matter is made of holes reveals a wealth of information about the universe. These waves created UPON the three dimensional surfaces WITHIN these three dimensional holes are responsible for the creation of a number of previously unexplained phenomena. Perhaps the most noteworthy are NEWTON'S THREE LAWS OF MOTION, MOMENTUM, and THE CONSERVATION OF MOMENTUM: Chapter 27.

Although many people think these laws have already been explained. This is wrong! They are listed as **Axioms**, meaning they have no explanation but have to be accepted as if they have been explained!

It should also be mentioned that the internal resistance to the creation of these waves is responsible for the creation of MASS and INERTIA. No Higgs Boson particle or Higgs field is needed.

Note: Even though a region of denser space surrounds an electron, when it is caught in the gravitational field of a star or planet, its three dimensional hole will distort exactly like the proton in the direction of the field. Consequently, it too will be accelerated towards the less dense space surrounding the star or planet.

Because quarks are fourth dimensional holes existing upon the surface of fifth dimensional space, it is much more difficult for three dimensional space to distort their surfaces [making it appear as if they are more massive]. However, because the fourth dimensional space within the three dimensional hole is itself slightly distorted by the distorted shape of the three dimensional hole, these fourth dimensional holes are slightly distorted too, causing them to also move in the direction of the distortion: Chapter 26.

It should also be mentioned that if the proton is within an object that is for example resting upon the surface of the Earth, the proton will continue to be distorted and will continue to try to accelerate towards the center of mass of the Earth (pressing it against the surface of the Earth). Again, MATTER IS NOT PULLED TOWARDS THE CENTER OF MASS OF A GRAVITATIONAL FIELD – BUT INSTEAD, PUSHES ITSELF TOWARDS THE CENTER OF MASS OF THE EARTH!

Finally, it should be noted that the distortion seen in #3 also applies to a certain extent to the photon. Even though the photon is constructed out of a dense region of space, a distortion is created in it too by the less dense space surrounding the sun.

This region of less dense space surrounding the Sun will be slightly greater on one side of the photon than the other. This will distort it ever so slightly in the direction of the Sun and make it move closer to the Sun as it flies past.

This distortion will create the famous effect of making the light seen from stars during a total eclipse of the Sun appear to move inward towards the Sun. It also created the "lens effect" [the bending of light] seen in deep space above and below galaxies.

> Contrary to present belief, there are not just four forces in nature but a fifth one too. Because it is opposite to the force of Gravity, this fifth force can be designated the **anti-gravity force**. This anti-gravity force is responsible for a number of unexplained phenomena that occur in nature. Someday, when this force is artificially created, and moves spaceships, it will not only be the stunning confirmation of the Vortex Theory, but also, it will be the greatest technological achievement ever made. It will allow us to finally travel to the stars, and will make atomic energy created by fission obsolete!

Chapter 18
The Discovery of the Fifth Force in Nature:
The Anti-gravity Force!!!

Because nothing is presently known about this force, the phenomenon that it is responsible for creating will be revealed. Incredibly, one of the most famous and well known effects the anti-gravity force is responsible for is Buoyancy!

We all know about the phenomenon called Buoyancy, this is the reason why wood and boats float in water; however, until now, no one knew the REAL CAUSE of this phenomenon!

BUOYANCY…

Buoyancy is important to us all. Without buoyancy, no ship could sail the sea, nor could anyone swim in the water. Buoyancy is another one of those phenomenon of nature that is so old its acceptance is without question. We know that it exists, and because it is a phenomenon that is so common, we don't even think to question it.

But why does buoyancy exist? What is the mechanism in nature responsible for making an object "buoyant"? What makes something float upon the surface of the water? What makes minerals of different densities separate apart from one another in a revolving fluid such as those in a centrifuge?

THE ANSWER IS SOMETHING NOBODY HAS EVER SUSPECTED BEFORE.

It is not the mass of the object (such as a piece of wood) that causes buoyancy, but the density of the space within the wood! It works like this: the *more* protons and neutrons per cubic inch, the *less dense* the space within. Subsequently, the *fewer* the protons and neutrons per cubic inch, the *denser* the space within. Therefore, it can be said that within a cubic inch of a mineral containing a lesser number of protons and neutrons than another cubic inch of a different mineral containing more protons and neutrons, regions of denser or less dense space are in effect "trapped." This relationship can be seen in Figure 18.1, A and B: the cross sections of rocks.

Figure 18.1

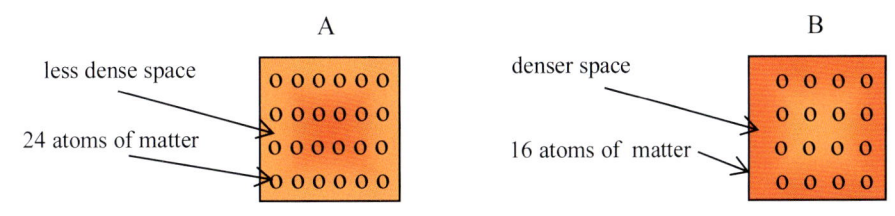

BUOYANCY IS OPPOSITE TO WHAT INTUITION TELLS US!

In Figure 18.1 A, **more matter** per cubic inch creates a region of **less dense space** in comparison to Figure 18.1 B. In B, **less matter** per cubic inch creates a region of **more dense space** in comparison to figure A. Within a stream of flowing water, when these different regions of space are being mixed together, or allowed to rotate together, the turbulence of the water allows them to rearrange their locations. The **more dense regions of space** found in **less massive** rocks try to move upward; while the **less dense regions of space** trapped in **more massive** rocks move downwards. In effect, *when moving*, these denser regions of space possess anti-gravity properties! This is opposite to what intuition tells us!

It must be emphasized that even though all of the protons and neutrons from both cubic regions of space are accelerated towards the Earth's center of mass equally, when mixed together, it is the denser region of space that is bent upwards and away from the direction of the Earth's center of mass. Consequently, this denser region of space that is trapped within a less dense rock seeks to move upwards and away from the Earth creating a buoyancy effect.

This same effect is produced within the hull of a ship. A region of more dense space is "trapped" within the empty hull of the ship making it push up out of the water.

A hot air balloon rises because *photons are dense regions of space*. When a massive amount of photons are trapped within an enclosure, such as a balloon, the interior region of air that is constantly exchanging them has in effect entrapped a volume of denser space. Because the surrounding space outside the balloon is less dense, the denser region of space inside of the balloon is accelerated upward.

THE UNIFORM DISPERSAL OF ONE GAS WITHIN ANOTHER GAS…

Another example of anti-gravity effects upon the earth that no-one knows about is the uniform dispersal of one gas within another.

In solids, atoms and molecules are in fixed positions. Between these atoms and molecules are trapped regions of dense or less dense space. It is these dense and less dense regions that create buoyancy. But within gasses, a different situation is occurring.

Within gases, there are no trapped regions of dense or less dense space. All of the atoms within the gas are pressed against one another. This means the electrons in the outer shells of the atoms are all pressed together but not bonded to each other. Because of this fact, the bent outward regions of space surrounding the electrons in the outer shells of the atoms slightly repel each other.

This repelling effect causes all of the atoms to move apart from one another, to disperse throughout the gas, and seek positions where the effect is at its minimum.

One of the most important questions that 20th Century science has never satisfactorily answered is how come atoms do not merge into other atoms? What keeps the nuclei of atoms from just piling up and creating one gigantic atom?

The answer is the denser space surrounding the electron. This region of denser space repels the denser space surrounding other electrons in other atoms and keeps them apart. In effect, Anti-Gravity keeps atoms from merging into each other. It also is responsible for the acceleration of galaxies away from each other as revealed by the Hubble Telescope

RESPONSIBLE FOR THE ACCELERATION OF THE GALAXIES AWAY FROM EACH OTHER!

In the late 1990's, a picture taken from the Hubble telescope seemed to indicate that a supernova in a Galaxy located in a distant part of the universe was accelerating away from us at a much faster rate than it should. But how can this be? How can all of the previous spectrographic analysis of the pictures from the last 50 years showing the Red Shift of the Galaxies indicate one result, and this picture indicate another?

The answer to this dilemma can be resolved when it is realized that this picture was taken on the other side of a region in the universe where no galaxies can be seen with earth bound telescopes. This indicates that there is a vast sea of space between the Galaxies in this region.

In these vast reaches of space where there are few Galaxies, the space is denser. Hence, it tends to cause all of the sub-atomic matter in the Galaxies located on either side of it to be distorted in the direction opposite to it, causing these Galaxies to accelerate away from this vast wasteland towards other less dense regions of space. But this effect is compounded by what is popularly called Dark Energy!

The Dark Energy effect is being created by massive clouds of Anti-hydrogen. This subject is explained in Chapter 31.

The discovery of how the photon is created, how it expands and contracts, finally allows us to explain one of the most perplexing and enduring mysteries in all of science - the particle and wave characteristics of light...

PART V
ENERGY

> Photons of energy are condensed packets of three dimensional space thrown from vortices. This condensed packet of 3d space both displaces the space surrounding it outward while it expands and contracts as it moves through the universe. The condensed region of space that is the photon creates the particle effect; the rotation or spin of photons creates their electromagnetic effect; their expansion and contraction creates their electric effect; while the region of denser space that is pushed outward and surrounds it as the photon passes through the universe, combined with the photon's expansions and contractions in the surrounding space, creates its wave effect!

Chapter 19
The True Explanation of the Photon

We all know that everywhere we look in nature we see an incredible variety of colors and shapes. However, what most of us don't know is that we never see anything! We don't even see the words on this page!

The only things we ever "see" are "photons" of energy.

Photons of energy are emitted from an energy source, hit the page, "bounce off", and hit our eyes. Without these photons we wouldn't be able to see anything at all. But just what are we "seeing"? What are these "photons" of energy? (What is energy?)

One of the greatest mysteries in all of physics is energy: what is it?

According to present day science, energy is contained within tiny particles called photons. These photons possess both particle and wave characteristics. But where they come from, what they are made of, or how they are constructed is unknown.

But not anymore!

With the discovery of the two vortices flowing back and forth between protons and electrons, the great mystery of the photon is finally unraveled - and a fascinating vision unfolds before us. What contemporary science calls a photon is really a condensed portion of three dimensional space that is thrown out of the vortex. It is too bad that 19th Century believers in the Aether theory got it backwards; they believed that matter was a "condensation" of space (like ice floating in water). If instead, they had said that energy was a "condensation" of space, they would have been much closer to the truth!

Although there are several ways to create photons, in the hydrogen atom, a photon is part of the volume of three dimensional space flowing from the electron to the proton. When certain conditions occur, the vortex shortens, throwing a packet of condensed three dimensional space back into the three dimensional space from whence it came.

The photon's velocity is also very important. The velocity of the photon is the velocity of the vortex, allowing us to deduce that the speed of light is a function of the speed of the vortex!

Although the present vision of the universe states that the photon "instantaneously" travels at the speed of light, we can now see that this is not true at all. It can readily be seen that a photon is but a part of the volume of three dimensional space in the vortex that was already circulating around and around at the speed of light before it was thrown free of the atom.

When a photon approaches an atom, is absorbed through the proton, and then is emitted through the electron. These principles can be seen in the following drawings:

Figure 19.1

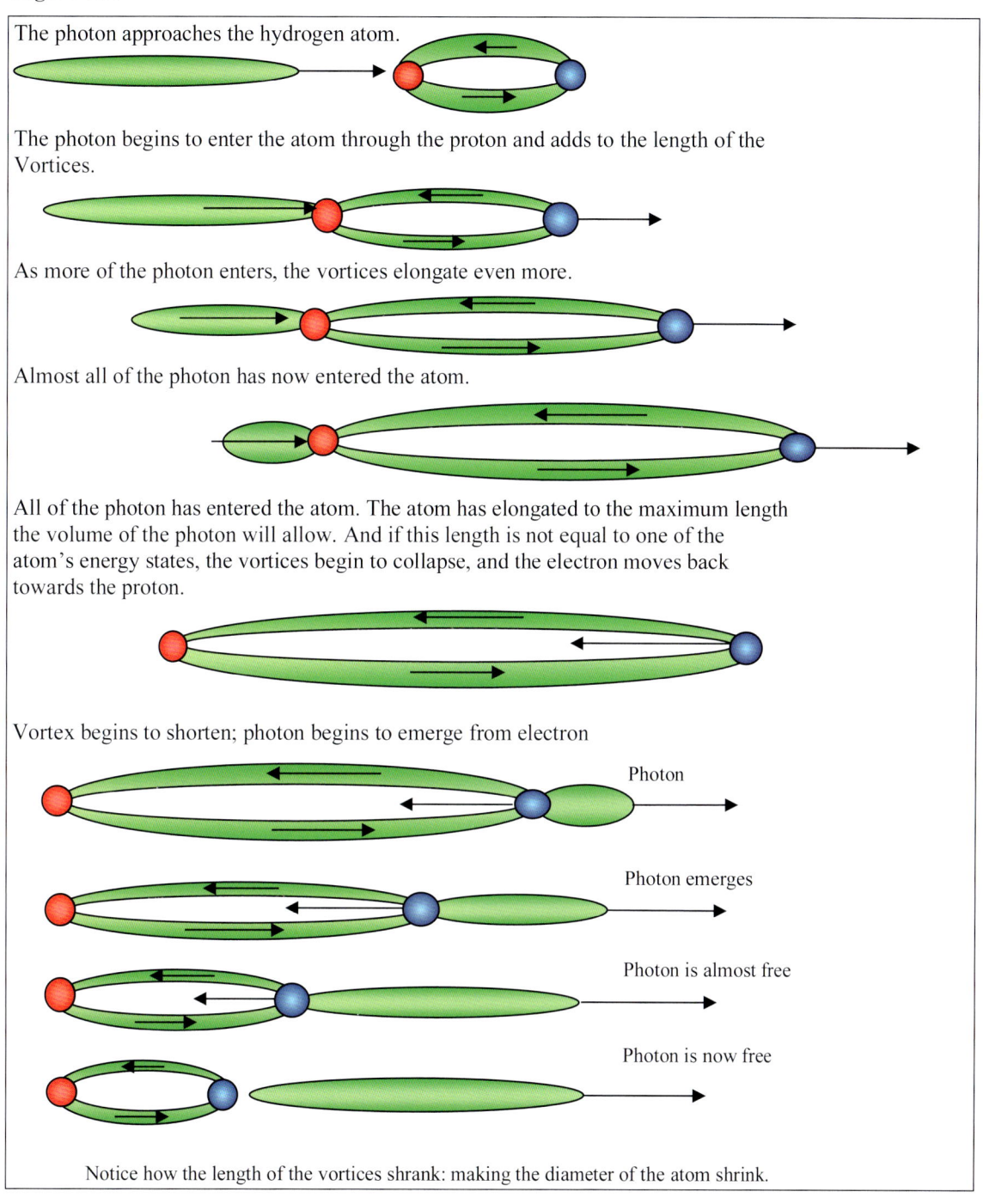

The photon approaches the hydrogen atom.

The photon begins to enter the atom through the proton and adds to the length of the Vortices.

As more of the photon enters, the vortices elongate even more.

Almost all of the photon has now entered the atom.

All of the photon has entered the atom. The atom has elongated to the maximum length the volume of the photon will allow. And if this length is not equal to one of the atom's energy states, the vortices begin to collapse, and the electron moves back towards the proton.

Vortex begins to shorten; photon begins to emerge from electron

Photon

Photon emerges

Photon is almost free

Photon is now free

Notice how the length of the vortices shrank: making the diameter of the atom shrink.

A PHOTON IS NOT "REFLECTED"; IT IS ABSORBED INTO THE VORTEX THEN RE-EMITTED!

Using the revolutionary principles of the Vortex Theory, we see that the photon is not "reflected" off an atom, but instead, is absorbed by one of the vortices of the atom and then re-emitted.

One of the most elegant and beautiful relationships in all of nature is revealed in the absorption and expulsion of the photon. Because both the photon and the vortex are moving at the speed of light, when the photon enters the flow of the vortex it does not interfere with the flow; and it exits without creating any interference. It is equivalent to a man stepping off of a moving escalator and onto a moving walkway going at the same speed.

These illustrations not only allow us to see how a photon is absorbed and then emitted from an atom, but also, allows one to explain many more of the previous heretofore mysteries of light reflection, refraction, and absorption.

One of the most elegant relationships in all of nature that is *not* revealed by these illustrations is the speed of the absorption and expulsion of the photon. Because both the photon and the vortex are moving at the speed of light, when the photon enters the flow of the vortex it does not interfere with the flow; and it exits without creating any interference to the flow. This relationship never changes, even the speed of the atom is irrelevant. Such a harmony is not only elegant it is also very beautiful.

Great beauty is also found in the wave aspects of the photon.

When a photon is emitted from an atom, it is an extremely dense packet of space. Consequently, it displaces the surrounding space, creating a dense region of space around it much like the space surrounding the electron:

Figure 19.2

The particle and wave aspects of a photon. Note, the sphere of dense space that surrounds the photon is massive. Because of space limitations on this page, this drawing is only a representation of the huge sphere of dense space surrounding the photon.

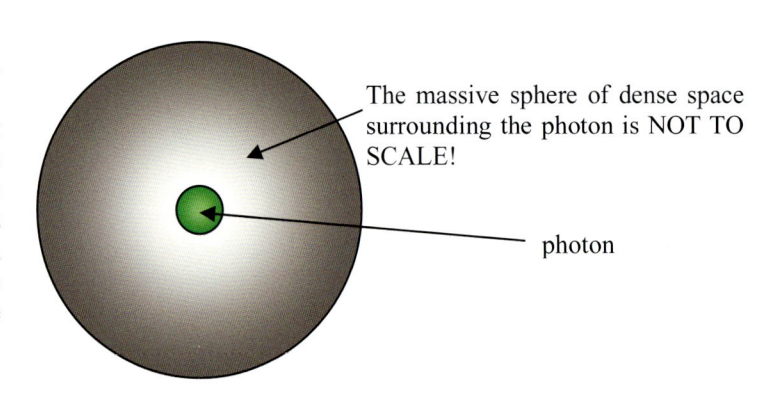

The massive sphere of dense space surrounding the photon is NOT TO SCALE!

photon

And again, just like the region of less dense space that surrounds the proton, the region of dense space that surrounds the photon is a function of the three dimensional space the photon is traveling through. This means that as the photon moves through the three dimensional space of our universe, the space the photon is passing through bends outwards away from the photon as it approaches, and then back inward as it passes by. Consequently, even though this massive sphere of dense space constantly surrounds the photon and seems to travel with it, *it is the dense region of space the photon is constructed out of that displaces the surrounding space, making it first become denser, then less denser again as it passes through.*

Using the principles of the Vortex Theory, the creation of the photon's electric and magnetic components are explained: the condensed region of space is responsible for creating the photon's electric component and its particle effect; its expansion and contraction is responsible for its frequency; its motion through three dimensional space creates a wave in the surrounding space. This wave is responsible for the photon's magnetic component and wave characteristics. The simultaneous expansion and contraction of both the dense region of space that is the photon and the surrounding space it passes through explains why the electric and magnetic effects are at right angles to each other. Also the photon's particle and wave characteristics are explained. This is made more apparent and illustrated next…

> Maxwell's differential equations are one of the great treasures of 19th Century physics, and they are treasured today. These equations describing the electromagnetic fields of light are considered to be a great achievement. Unfortunately, Maxwell made a great mistake, he described the effect and not the cause!

Chapter 20
Maxwell's Mistake: He Described the Effect and NOT the Cause!

In 1865, James Clerk Maxwell proposed that light was an electromagnetic wave consisting of an oscillating electric and a magnetic field. And as the electric field expanded, the magnetic field contracted and vice versa. It must be said that this was a brilliant idea, but Maxwell had no idea how these two fields were being generated.

Unfortunately, this great idea was a mistake! His brilliant differential equations explain the effect but not the cause!

Mr. Maxwell did not know how the photon was constructed. If he did, he would have realized that his 'electromagnetic wave" was an effect and not a cause. The true cause for the creation of what he called the electric and magnetic fields is the photon's denser space trying to merge back into the surrounding space. Finding out that it cannot and is forced to expand and contract in directions perpendicular to the direction of travel. And as it does, its subsequent expansion and contraction combined with its rotation causes the surrounding space to rotate in an ever increasing diameter only to reach a point where, when the compressed space directly above and below the photon causes it to contract back upon itself: causing the rotation of the surrounding space to expand outward. This concept can now be drawn…

In the drawings below, it can be seen that when the dense space of the photon expands and contracts perpendicular to the direction of travel, the expansion and contraction of its magnetic field occurs parallel to the direction of travel: making the angle between the electric and magnetic fields occur at ninety degree angles to each other.

Also in the below drawings, note in Figure 20.1.1 & 20.1.5 how the density of the surrounding space creating the magnetic force is weak as the photon reaches its maximum extension; vs. that as seen in Figure 20.1.4 & 20.1.8 when the photon has now reached its maximum compression.

Notice how in Figure 20.1.2 & 20.1.6 how the magnetic field begins to expand when the extended photon begins to contract and shrink in length. Note too in Figure 20.1.4 & 20.1.8, when the photon has reached its maximum contraction, how the surrounding magnetic field is at its maximum extension.

Figure 20.1

Figure 20.1.1 **Figure 20.1.2** **Figure 20.1.3** **Figure 20.1.4**

Note: sizes of magnetic and electrostatic component are <u>not</u> proportional; magnetic component is enormous compared to size of electrostatic component.

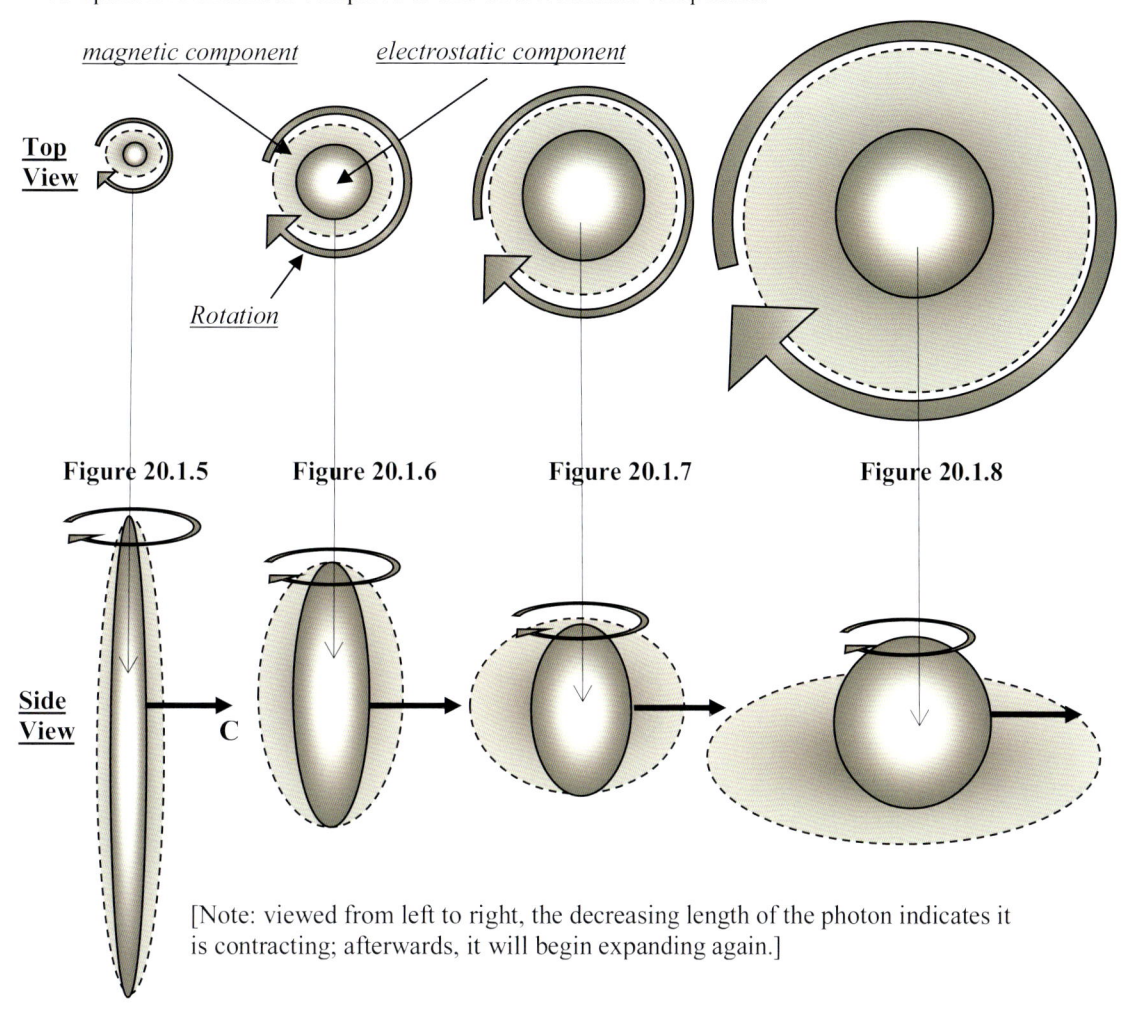

magnetic component   *electrostatic component*

<u>Top View</u>

Rotation

Figure 20.1.5 **Figure 20.1.6** **Figure 20.1.7** **Figure 20.1.8**

<u>Side View</u>

C

[Note: viewed from left to right, the decreasing length of the photon indicates it is contracting; afterwards, it will begin expanding again.]

20th CENTURY PHYSICS AND THE ROTATION OF THE PHOTON?

Although 20th Century science presently believes that the photon has to rotate in a direction parallel to its velocity of travel, this is a mistake. This idea is based upon the mistaken premise that to rotate perpendicular to the velocity of travel causes it to move faster than the speed of light: a condition that is impossible according to 20th Century beliefs and reasoning. However, this incorrect deduction is easily dispelled: the increased density of the more massive photons creates a greater density in the surrounding space, allowing it to vibrate faster; making it only seem from our perspective to move faster than the speed of light. In reality, the denser space surrounding it allows it to rotate faster.

Figure 20.2 **Figure 20.3**

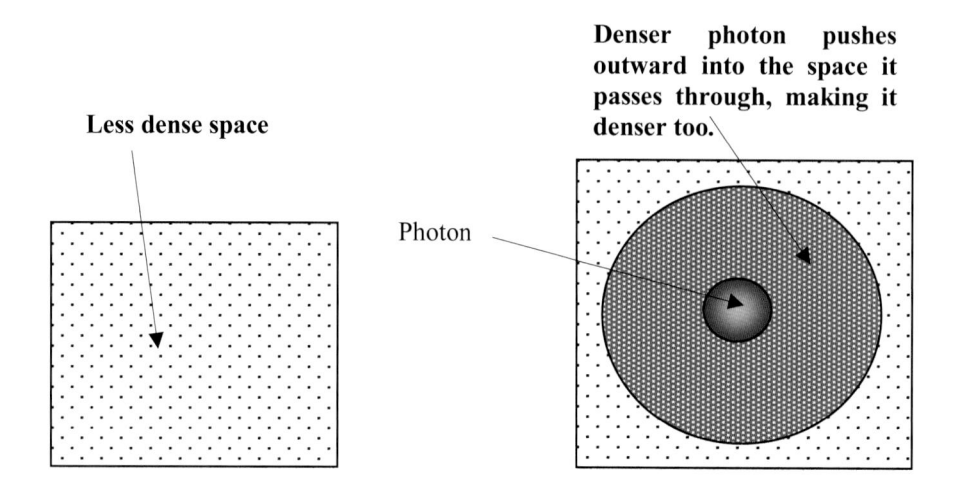

Less dense space

Denser photon pushes outward into the space it passes through, making it denser too.

Photon

Consider the gamma ray. It is easily demonstrated through vector addition that this powerful photon with a frequency of 10^{19} to 10^{30} is already expanding and contracting faster than the speed of light. But again, *the increased density of the more massive photons creates a greater density in the surrounding space*, allowing it to vibrate faster: only making it seem to move faster than the speed of light from our perspective, as mathematically explained in Figure 20.4 below for those interested in mathematics...

Figure 20.4 For science...

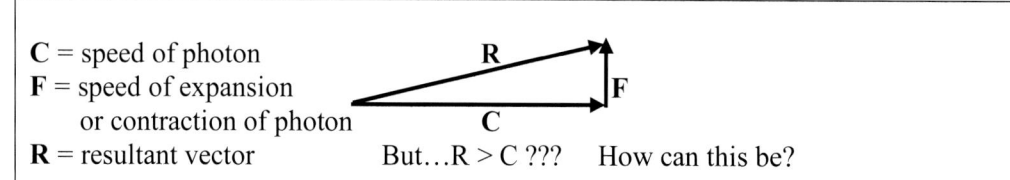

C = speed of photon
F = speed of expansion
   or contraction of photon
R = resultant vector   But...R > C ???   How can this be?

How can this be indeed! The explanation for this apparent violation of universal law is the density of the space surrounding the photon. A more massive photon is surrounded by a larger more massive region of dense space. This region of dense space allows the photon to vibrate faster. If measurements of the speed of light could be taken in such a region of denser space, it would reveal that everything is moving faster, including the electrons in the instruments making measurements [including atomic clocks], and no increase in anything especially the speed of light would be apparent!

Because the resultant vector represents a velocity that is greater than the speed of light, the argument that nothing can move faster than the speed of light is erroneous.

Also, it is important to note that the perpendicular rotation of the photon in no way affects the angular momentum it imparts to an object. It is important to remember that the angular momentum of particles is quantized. Whenever a photon is absorbed by an object, its angular momentum of either $+h/2\square$ or $-h/2\square$ is quantized. Hence, the value of the "quantized angular momentum" is still imparted to the object no matter if it is spinning parallel or perpendicular to the object.

THE PHOTON'S VELOCITY IS A FUNCTION OF THE SPEED OF THE VORTEX

The creation of photons is very important because they reveal that they are thrown from a vortex or created by the interaction of magnetic fields or electrostatic fields: all of which are moving at the speed of light. This is extremely important information because it reveals that the velocity of the photon is the velocity of the speed of the vortex! Allowing us to make a most profound deduction: *the speed of light is a function of the speed of the vortex!*

Although the present vision of the universe states that the photon "instantaneously" travels at the speed of light, we can now see that this is not true at all. It can readily be seen that a photon is but a part of the volume of three dimensional space that was already circulating at the speed of light before it was thrown free of the atom.

The creation of the photon's *Electromagnetic characteristic* is a result of its spin. Because the electron is spinning when the photon is thrown from the vortex it is spinning too;

Figure 20.5 below represents two photons seen edge on and moving from left to right across the page. The first photon A; is spinning clockwise; while photon B is spinning counterclockwise. They are both traveling at velocity v, the speed of the vortex from which they were thrown. We recognize this velocity as "C", the speed of light. But it must be remembered that C is in reality merely a function of the velocity of the vortices speeding back and forth between protons and electrons.

Figure 20.5

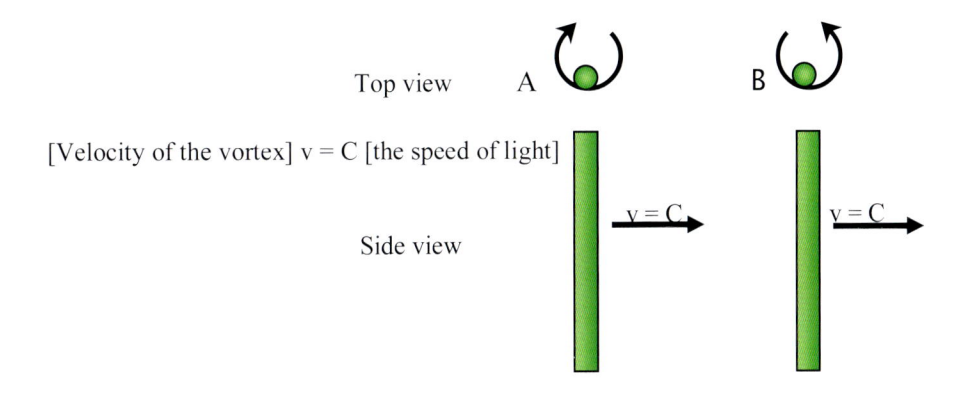

Top view   A   B

[Velocity of the vortex] v = C [the speed of light]

Side view   v = C   v = C

Because both photons are expanding and contracting perpendicular to the direction of travel, it is interesting to note that they can only have one of two spin orientations: clockwise or counterclockwise! The Perpendicular Expansions and Contractions of the Photon are a most ingenious characteristic of nature.

The photon has to expand and contract outward in long tubular shapes rather than flattened out pancake shapes. If it expanded and contracted in a flattened out "pancake" shape, the velocity of space in front of it would have to exceed the speed of light to get around it, creating a "Rip" or "Tear" in the surface of three dimensional space.

Even though the speed of the expansions and contractions of the photon slightly exceed the speed of light, the region of denser space that surrounds it allows this effect. Denser space has a higher elastic modulus that allows it to bend and flex faster.

The reason why the photon even expands and contracts, is due to its attempt to blend back into the three dimensional space from which it came. As the photon races away from the electron it was emitted from, it tries to expand back into the three dimensional space it originally came from but instead finds that it can only expand in a direction perpendicular to its velocity of travel. If it tried to expand parallel to its direction to travel, its forward direction of expansion would exceed the speed of light, creating a tear in space. Or, if it tries to expand perpendicular to the speed of space in a pancake shape, the space would have to exceed the speed of light to get around it. Hence, it must expand in a perpendicular tubular direction.

The expansion of this dense region takes place in a tubular shape allowing the space in front of it to swiftly move around it. As it expands upward and downward simultaneously, the space immediately above and below it is pressed upward and downward respectively. It is the pressure of this surrounding space along the y axis that keeps the photon from expanding back into the three dimensional space it is made of. This event occurs when the photon, having expanded as far as its elasticity allows it, finds that the space immediately above and below it is now pushing back down upon it, causing it to contract back upon itself.

This contraction continues until the photon is forced back into the condensed spherical shape, beginning its expansion all over again.

As a photon travels through three dimensional space, its expansion and contraction perpendicular to the direction of its velocity creates its frequency. The length of the beginning and ending of these expansions and contractions creates its wavelength.

Figure 20.6 Photon moving from left to right across page

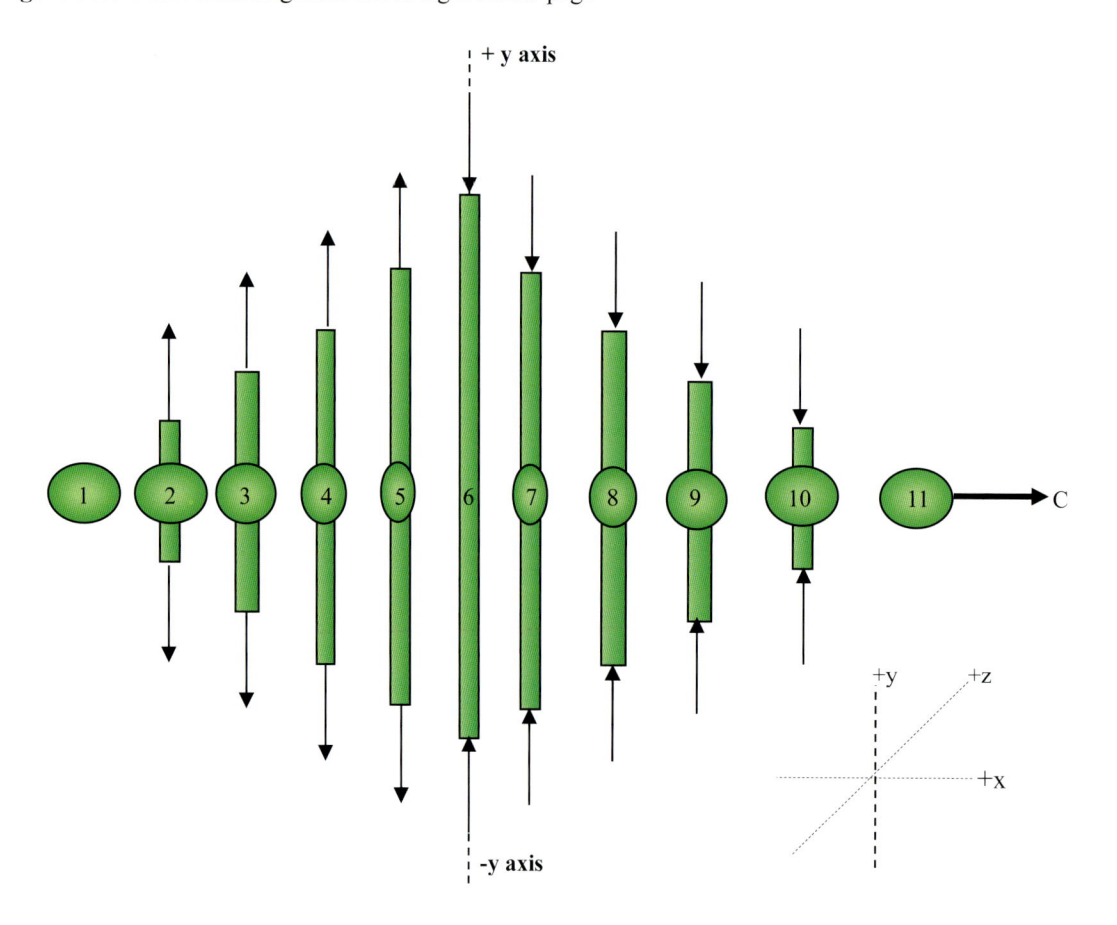

The volume of the denser space within the photon determines its length of expansion and hence its frequency. [Note: the expansion is in the ±y component of the Cartesian Co-ordinate system **and is caused by the dense space within the photon trying to blend back into the surrounding space.**]

The reason for the seemingly oscillating back and forth exchange of the electric and magnetic fields is a result of the tradeoff between the changing diameter of the photon. When the photon is in shape 1, it is rotating, and its radius is creating maximum rotation of the surrounding space. The magnetic field is <u>**at a 90 degree angle**</u> to the expanding and contracting electric field vibrations. **[NOTE: again, the magnetic field is caused by the rotation of the surrounding space created by the rotating photon; while its electric component is created by the expansion and contraction of the photon perpendicular to the direction of travel.]**

As the photon expands outwards upon the y axis, it can be seen that its diameter grows smaller and smaller, until at 6, the radius is very tiny creating only a very small amount of rotation in surrounding space. However, as it then contracts and goes from 7 to 11, it can be seen that as the photon contracts, its radius again continually increases, allowing it to again create maximum rotation in surrounding space.

But at the same time, it can be seen that as the radius of the photon grows smaller, the height of the expansion increases along the y axis. This expansion creates the electric effect of the photon. At #1, its height is at minimum while the magnetic effect is at maximum. While at #6, the magnetic effect is at minimum while the electric effect is at its maximum: visually allowing us to now physically observe Maxwell's brilliant deduction.

Note too: it should also be mentioned that the gamma ray possesses so much condensed space that when <u>*seen head on*</u>, it expands and contracts in a six sided "star pattern":

Figure 20.7

Figure 20.8 Seen sideways, it looks like this:

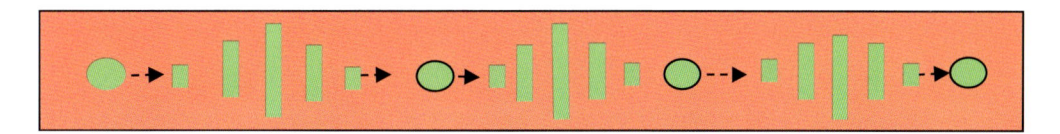

So to reiterate, photons of energy are condensed packets of three dimensional space that displace the surrounding three dimensional space outward, creating a spherical region of dense space that surrounds the photon. The expansions and contractions of the volume of space within the photon creates the *electric effect*. While the rotating region of denser space expanding and contracting perpendicular to its diameter creates its *wave effect*.

The photon can now be defined as a packet of condensed three dimensional space, expanding and contracting in a long tubular shape perpendicular to its direction of travel. The rate of its expansion and contraction - or frequency - is a direct function of the *volume* of condensed space within the photon.

It is also important to mention that the incredible fast vibrations of gamma rays and x-rays seem to go faster than the speed of light. This might seem to be true from our point of view. However, it must also be realized that these incredibly dense photons also create an incredibly dense region of space surrounding them. This denser space allows them to vibrate faster without creating rips in space. If we could go into this denser space and measure their vibrations, even though we cannot measure the effect because the electronics in our instruments in this denser space would be moving faster, we would not be able to see that the speed of everything appears to be moving faster than the speed of light. Hence the fast vibrations of gamma rays and x-rays are not exceeding the speed of light in this denser region of space.

MAGNETIC FIELDS CREATING ELECTRIC FILEDS?

There is no doubt that James Clark Maxwell was a brilliant scientist. He proposed that changing electric and magnetic fields should trigger each other and these changing fields should move at a speed equal to the speed of light. Concluding this reasoning process, Maxwell then stated that light *is* an electromagnetic wave; and later experiments confirmed Maxwell's theory." However, here is where the problem develops.

Maxwell died in 1879 and the electron was not discovered by Thompson until 1897. So up until 1897 nobody knew how the changing magnetic field was creating the electric field and vice versa. All that anybody knew was that the movement of a wire through a magnetic field or the placement of a wire in a changing magnetic field was generating an electrostatic field.

But that all changed after the discovery of the electron. With the discovery of the electron also came the realization that the electron was generating the electrostatic field. Furthermore, that it was the movement of electrons along the wire within the changing electromagnetic field that was creating the electrostatic field; and that the electron and its electrostatic field cannot be separated. Hence, an electromagnetic field cannot generate an electrostatic field. It can only move electrons that in turn generate the electrostatic field. However, and most fascinating is the opposite scenario.

An electrostatic field can generate a magnetic field. This effect can be traced to the spin of the electrons. As electrons move along a wire, their rotations create corresponding rotations in the surrounding space creating electromagnetic fields. [According to the vortex theory the magnetic fields of electrons are created by the rotation of space surrounding them; while electro-static fields are created by the space flowing out of the electron.]

When all of the individual rotations surrounding all of the electrons combine, they form the electromagnetic field about the wire. It is important to note that the strength of the magnetic field is proportional to the current in the wire making it proportional to the number of electrons moving along the wire.

So even though Maxwell was right that changing electrostatic fields can create changing electromagnetic fields, the opposite is not true. His failure came from his ignorance of the existence of the electron that was the cause of the effect he was seeing.

Consequently, it can now be seen that his explanation of the electrostatic field creating an electromagnetic field and vice versa in the photon is a mistake!

THE ORIENTATION OF THE PHOTON'S VIBRATION PROFILE CAN CHANGE!

The oncoming profile of the photon does not have to maintain any one particular orientation. In the following figure the oncoming picture of the photon can change from a pure x, y orientation to any other…[Note: the perspective shows the photon coming directly out of the page towards the reader in the +z direction.]

Figure 20.9

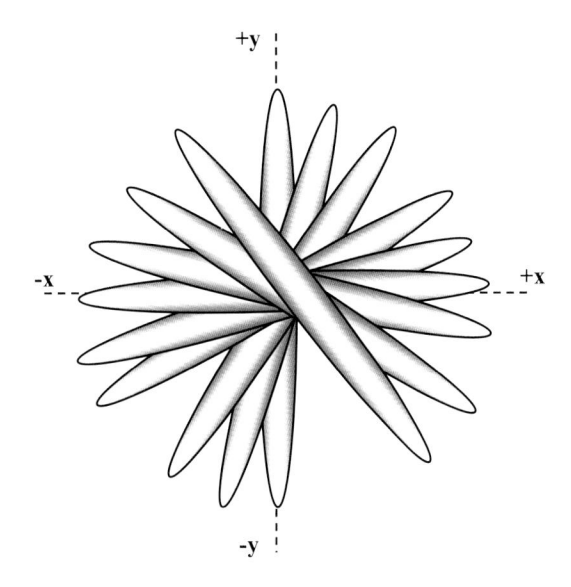

IN CONCLUSION

Using the principles of the vortex theory, the explanation of the photon's characteristics is easily explained. The photon's particle effect is created by a dense region of space emitted from a vortex at the speed of the vortex flowing at the speed of light; the wave effects of the photon are created by the space it is moving through; while its electric effects are created by the expansion and contraction of the photon; and its magnetic effects are created by the surrounding space. The rotation of the photon causes a corresponding rotation of the surrounding space

> Using the principles of the strong force, the Alpha particle is easily explained.

Chapter 21
The Alpha Particle

One of the great observation achievements of early particle scientists was the discovery that only three types of particles are expelled from the nucleus of an atom: alpha particles, beta particles, and gamma rays.

Now the gamma rays don't present a problem because they are nothing more than energetic photons; nor do the beta particles because they are merely electrons. However, the alpha particles –which are nothing more than helium nuclei – do present a very special problem: why only alpha particles? Why are only alpha particles thrown out of the nucleus? Alpha particles consist of two protons and two neutrons. So what is so special about this arrangement? Why don't we see particles made up of one proton and one neutron, or three protons and three neutrons, or four protons and four neutrons, or more?

The answer is found in the true nature of the strong force. The strong force is not an attraction between particles, but rather the continual transformation of a proton into a neutron and a neutron back into a proton as explained in Chapter 14. This process is easy to observe between particles that are paired with each other, but it is also easy to observe in two pairs of particles.

When a proton neutron pair is close to another proton neutron pair, the continual transformation keeps all four particles "stuck" together:

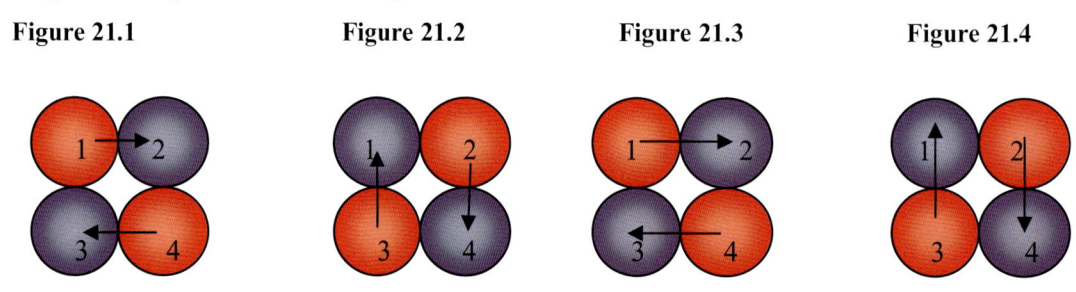

Figure 21.1   **Figure 21.2**   **Figure 21.3**   **Figure 21.4**

[note: red = proton: lavender = neutron]

In the above sequence note how 1 becomes 2, 2 becomes 4, 4 becomes 3, and 3 becomes 1 completing the circuit. Note also, that if 1 became 2 as 4 became 3, and then 2 became 1 again as 3 became 4 again each pair would be acting as a separate set instead of one set. In this circumstance, set 1&2 and 3&4 would be separate from each other and would not be held together by the strong force. Instead, the two sets would be pressed together inside of the nucleus due to the bent inward regions of space that surround each particle.

> Using the principles of the Vortex Theory, the construction of the alpha particle, and the theory that the nucleus is constructed out of alpha particles, the explanation of the Pauli Exclusion Principle is explained. If protons and electrons are connected to each other via fourth dimensional vortices, they spin in opposite directions. Since the alpha particle possesses two protons possessing opposite spins, their electrons also possess opposite spins. With a nucleus constructed out of alpha particles, all paired electrons in shells and sub-shells will spin in opposite directions.

Chapter 22
The Pauli Exclusion Principle

In 1925, Wolfgang Pauli, after a long and arduous effort to explain the so-called "anomalous Zeeman effect" finally discovered what has come to be known today as the Pauli Exclusion Principle. This principle states that no two electrons in the same atom can have identical values for all four of their quantum numbers: explained in Chapter 34.

> For example: If two electrons occupy the 1s orbital, the first three quantum numbers of both electrons are identical: $n = 1$, $\ell = 0$, and $m_\ell = 0$. Since all three of these numbers are the same, the Pauli Exclusion Principle states that their fourth quantum number, their spin number, has to be different: one electron must have a spin of $m_s = +1/2$, and the other have a spin of $m_s = -1/2$. Consequently, what the exclusion principle reveals is that two electrons in the same orbital must have opposite spins.

Unfortunately, although this brilliant deduction by Mr. Pauli explained the Zeeman Effect and eventually won him the Nobel Prize in physics, there has always been one major problem with this great discovery – it is an empirical relationship.

An empirical relationship is based upon observation rather than theory; there is no theoretical reason to believe in the relationship, only data reveals it to be so. And so it is with the case of the Pauli Exclusion Principle. For the past 80 years, nobody including Pauli himself knew what was creating it. It has been an observation without an explanation. Only now, with the discovery of the Vortex Theory does the answer finally become apparent.

THE FOURTH DIMENSIONAL VORTICES EXPLAIN THE PAULI EXCLUSION PRINCIPLE

The explanation of the Pauli Exclusion Principle is revealed by the fourth dimensional vortices existing between protons and electrons.

Remembering that the "particles" we call protons and electrons are really three dimensional holes connected to each other via a fourth dimensional vortex. Recalling that 3d space is bent into the proton [and the neutron] and out of the electron; that space is *less* dense surrounding the proton [and the neutron] and *denser* around the electron; and that three dimensional space flows into the proton [creating its electrostatic charge], through the vortex, then exits at the electron creating its opposite electrostatic charge, creates the following…

Figure 22.1

The proton and the electron are connected by a vortex of three dimensional space flowing from the proton to the electron in higher dimensional space!

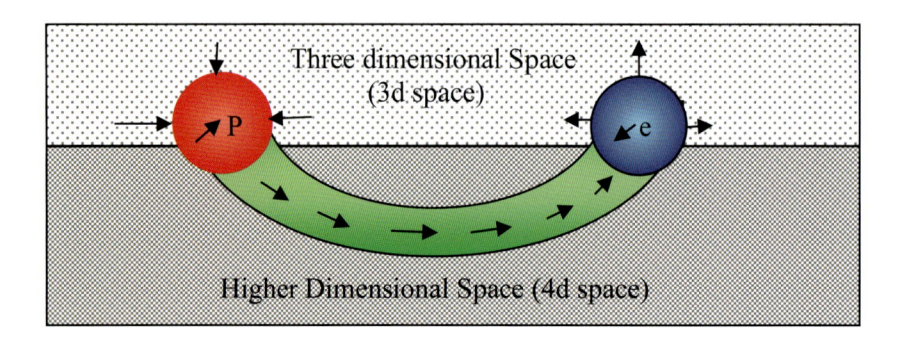

When a hydrogen atom is created, some of the space flowing out of the electron begins to flow into the proton; as these two holes move closer together a critical distance is reached where all of the 3d space flowing out of the electron flows directly into the proton. When this situation occurs, *a second* vortex of whirling space is created. These two vortices create a circulating flow containing a fixed volume of space.

This circulating volume of 3d space continually flows from the proton, into 4d space- through 4d space, and then into the electron. Here, it exits the electron, flowing back through 3d space and into the proton once again, binding the proton to the electron creating a hydrogen atom.

Figure 22.2

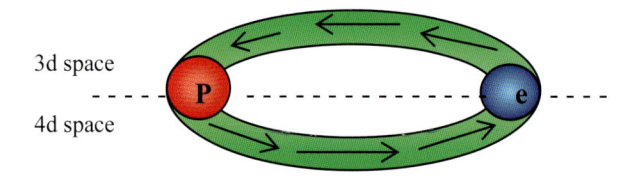

When the circulating flow commences, both of the electrostatic charges are neutralized. The word "neutralized" was used because no flowing space escapes from the system. If surrounding space still flowed into or out of this system, all atoms would possess electrical charges (and every time we touched something we would get shocked). Note: ions are created when a molecule is broken up and a proton in one atom is connected to an electron in another atom via a 4d vortex.

[Also: it is necessary to mention that the neutron is a vortex caught in a loop. The neutron is a proton completely surrounded by an electron: a hole within a hole. As such, it is turned into a 4d torus [a 4d vortex turned inside-out (similar to a smoke ring though impossible to draw)].

A CURIOUS OBSERVATION

The bases of the Vortex Theory now allow us to explain the Pauli Exclusion Principle. An explanation that begins with a most curious observation regarding the nucleus of the atom: why does the nucleus emit alpha particles? The alpha particle is a helium nucleus containing two protons and two neutrons: but why this combination? Why only two protons and two neutrons?

Why aren't particles emitted that contain three protons and three neutrons; or four protons and four neutrons? Also why is the number of protons and neutrons equal? Why aren't particles ejected that contain two protons and one neutron; or two neutrons and one proton?

THE ALPHA PARTICLES

The answer to the above question is found in the unique structure of the helium nuclei. This uniqueness is first found and characterized by its neutral spin. This neutral value reveals that the spins of all four nucleons cancel. And if the vortices exist, because the spin states of the electrons in a helium atom are opposites, the spin states of the protons are opposites; also, to create neutrality, the spin states of the neutrons must cancel; creating an effect similar to tiny magnets all holding each other together. It is this neutrality that is holding the alpha particle together.

According to the Vortex Theory, the intrinsic magnetism of fermions is caused by the rotation of the space around them; [fermions are spinning around a fourth dimensional axis]. Opposite spin states create opposite rotations creating opposite polarities. Because the 2 protons and the 2 neutrons in the alpha particle possess opposite spins they possess opposite polarities. These opposite polarities cause the 2 protons and the 2 neutrons to be attracted to each other much like *Cooper pairs*; binding one proton neutron pair to the other – binding the alpha particle together.

The alpha particles in the nucleus of the atom are bound together by the less dense space surrounding protons and neutrons. According to the Vortex Theory, the accumulation of all the less dense volumes of space surrounding all the protons and neutrons in a planet or star is responsible for creating its gravitational field. On the subatomic scale, these same volumes of less dense space are responsible for creating "nuclear gravity" that bind the alpha particles together in the nucleus, along with the excess neutrons responsible for creating isotopes of a particular atom.

Surprisingly, using the principles of the Vortex Theory, the explanation for Pauli Exclusion Principle can now be illustrated; [although the spin states of each type of particle is opposite, note particularly the spin states of the electrons]:

Figure 22.3

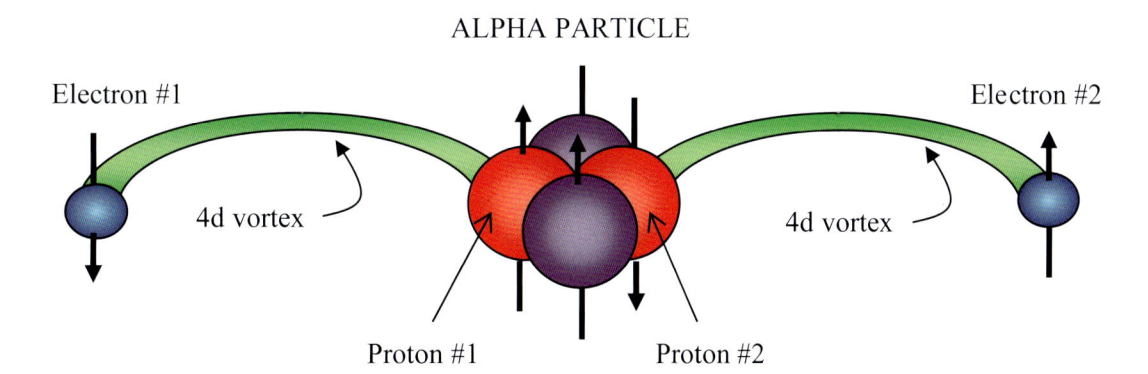

ALPHA PARTICLE

Electron #1   Electron #2

4d vortex   4d vortex

Proton #1   Proton #2

Note how the spin states, [the directions of the arrows], of the protons in the alpha particle are opposite to each other. Although the spin states of the proton and the electron it is attached to are also opposites; because each is at the other end of the rotating vortex, the spin states of the two electrons are now opposite to each other; consequently, *the spin states of the two electrons are opposite to each other because the spin states of the protons they are attached to are opposite to each other.* This is the simple yet elegant explanation for the Pauli Exclusion Principle.

Figure 22.4

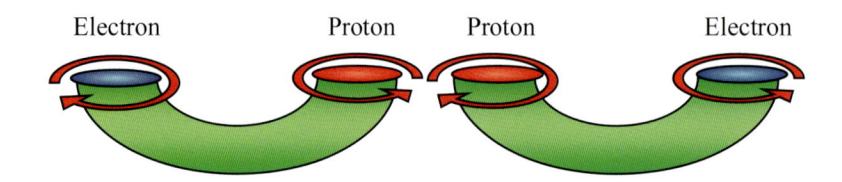

> We have just seen how photons of energy are expelled out of the electrons in an atom. However, more powerful photons called Gamma rays and X-rays appear to be emitted from directly out of the nuclei of atoms. How can this be? The following fascinating description and drawings explain it all.

Chapter 23
Gamma Rays and X-Rays

One of the great mysteries of how atoms absorb photons then expel them was solved many years ago and explained in the second book of this three part series, *The Vortex Theory of Atomic Particles*. Here it was discovered that photons are absorbed by the proton in say a hydrogen atom. Then the photon travels through the fourth dimensional vortex and is expelled out of the electron and back into three dimensional space. However, if it possesses just the right volume of three dimensional space, it can lengthen the vortex until the electron reaches a higher energy state [a greater diameter]. Here the electron stays until the atom finally expels the added volume of space in the form of a photon possessing the exact same volume of space the original photon possessed. But there is something else that can happen too! Something strange!

In some larger more massive atoms such as uranium [U238], a powerful photon called a gamma ray or an x-ray, can be expelled directly out of the nucleus of the atom! Many years ago, this did not seem to make any sense, and lent doubt as to the correctness of this vortex theory. Because if the model of the atom created by the Vortex Theory of Atomic Particles is right, the photon should enter the proton and exit through the electron! So how can it exit through a nucleus?

Although it was probably possible to answer this question many years ago, it was not until discovering the answer to the Constant of Fine Structure that it was realized how a more massive atom could expel a photon through its nucleus. It was a wonderful and satisfying discovery. It ended many years of doubt about the veracity of the Vortex Theory of Atomic Particles. It also solved the problem that persisted without an answer for over 20 years. This wonderful explanation all began with Schrödinger's equation and his placement of electrons in miniature "clouds" called "shells", and sub-shells".

Schrödinger's discovery was a great advancement to the science of physics and chemistry. But his vision of the atom was incomplete. What Schrödinger did not know about was the existence of the vortices. His electrons moved about within their shells connected to the nucleus of the atom by static charges. However, when his brilliant idea of shells *and sub-shells* are combined with the Vortex Theory of Atomic Particle's *two vortices*, suddenly, the explanation is revealed why some heavier elements such as uranium U238 emit radiation directly out of their nuclei: specifically alpha particles, gamma rays, and x-rays! It is ingenious and happens like this…

As the electron moves about in its shell, and gets closer to the nucleus and the proton to which it is attached to via its vortices, the vortices shrink in length and expand in diameter. The closer the electron approaches the nucleus, the thicker the vortices become; and the volumes of the proton and the electron have to both expand to be able to absorb and expel the extra volume. As the electron moves away from the nucleus, the vortices again lengthen out and the volumes of the proton and the electron contract; see Figure 23.1 below…

Figure 23.1

Part 1 The vortices expand and contract in size causing the proton and the electron to do the same

[for the ease of understanding, the proton and electron are shown the same size]

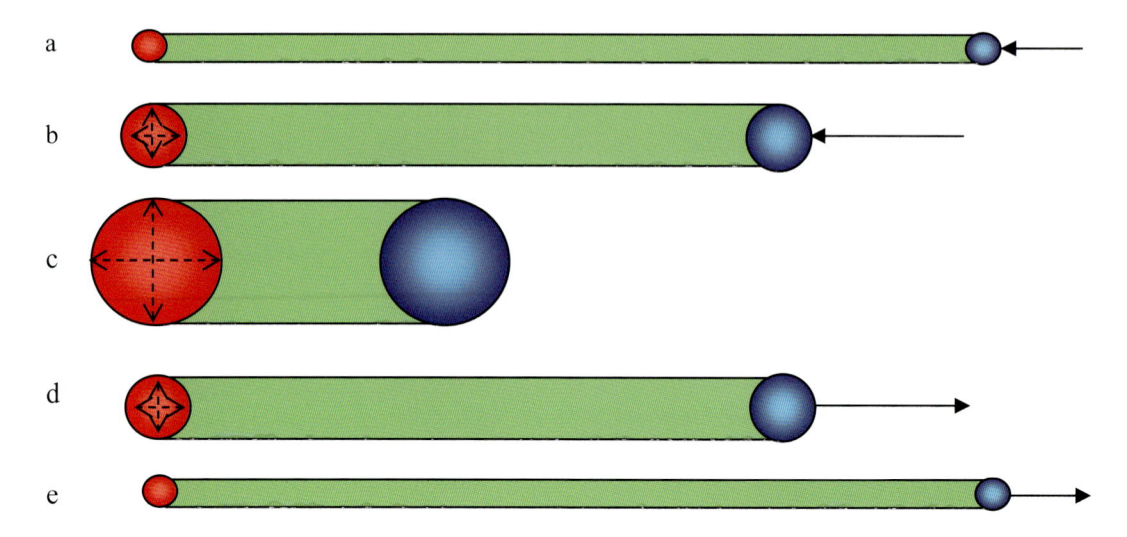

a

b

c

d

e

Now if the electrons in the P orbitals of the 2n shells [see Figure 23.2] all approach the nucleus of the atom at the exact same time, the vortices between them and the protons within the nucleus they are connected to will suddenly expand in size [see Figure 23.1, c above], causing the proton's volume to also expand in size…

Figure 23.2 P orbitals of the 2n shells

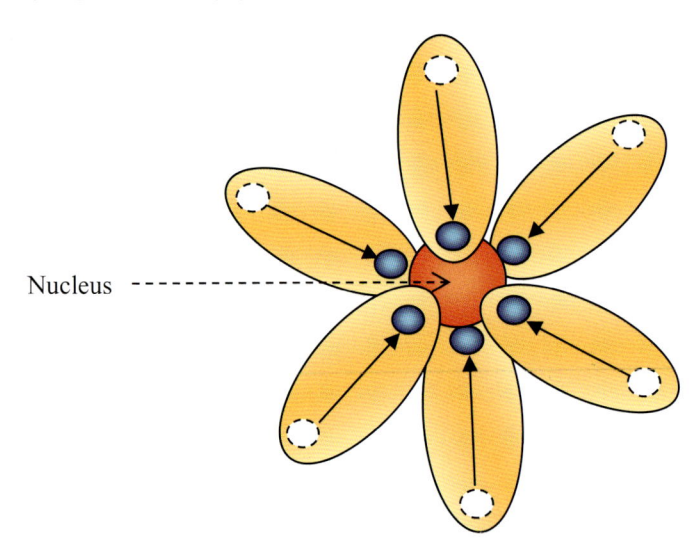

Nucleus

Part 2 The expansion and contraction of the proton creates a pressure wave within the nucleus

If an alpha particle is in the center of a nucleus and is being tightly pressed on all sides by other surrounded alpha particles making it difficult to move or change positions; and if a point in time is reached where all six of the protons in the surrounding alpha particles [one for each side of the alpha particle] all expand together at the same time in phase creating a resonance; then the less dense space surrounding them and creating their nuclear gravity, generates a pressure wave [grey sphere] that pushes inward, into the alpha particle; causing this alpha particle they surround to be suddenly compressed.

Figure 23.3

[Note: the expanding protons are actually much closer together than shown here; they are only drawn further away to allow the arrows to be inserted.] Note too: the thick black arrows represent the expanding volumes of less dense space surrounding the protons that create their nuclear gravity. This creates a gravitational wave [seen in grey] that suddenly compresses the alpha particle…

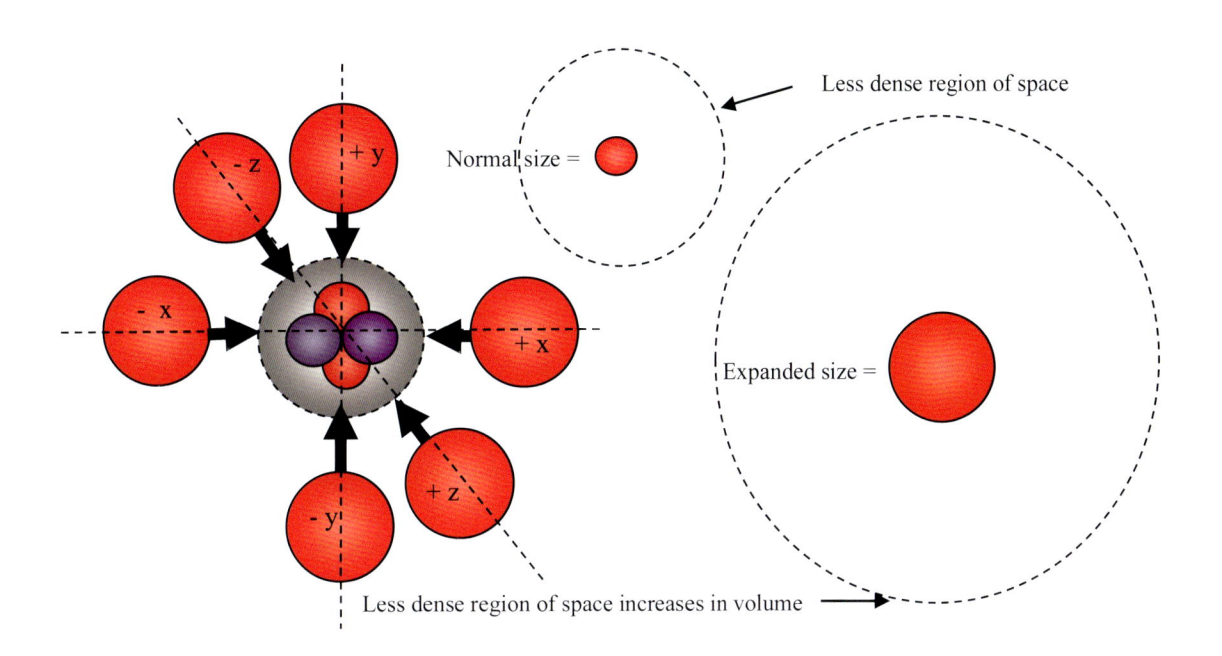

This sudden external pressure on the alpha particle causes it to be "pushed" out of the nucleus and into higher dimensional space: causing it to "tunnel" – (to move out of 3d space into 4d space); see Figure 23.4 below…

Part 3 The creation of the alpha particle and the gamma ray

When the alpha particle is suddenly compressed, it is thrown free of the nucleus and enters 4d space. Then, [all the following happens at the same time], since no 3d space flows into the two protons in 4d space; the two three dimensional vortices connecting the two electrons to the two protons in the alpha particle continue onwards to where the protons once were; their three dimensional volumes are shot into the void where the alpha particle was and then out through the nucleus and into the surrounding 3d space, becoming either gamma rays or x-rays.

If the vortices are long, their volumes will be large and they will form gamma rays; if the vortices are short, their volumes will be smaller, and they will form x-rays: the alpha particle is also shown here; it first enters 4d space at the speed of the compression wave created by the surrounding alpha particles. It then returns back into 3d space. When it does, the inward flow of the surrounding 3d space into the two protons recreates its "charge"; that in turn, flows back into the electrons via the 4d vortices, reestablishing their outward flowing "charges".

[Note: the 3d vortices in red represent the expulsion of their volumes that have now become the gamma rays.]

Figure 23.4

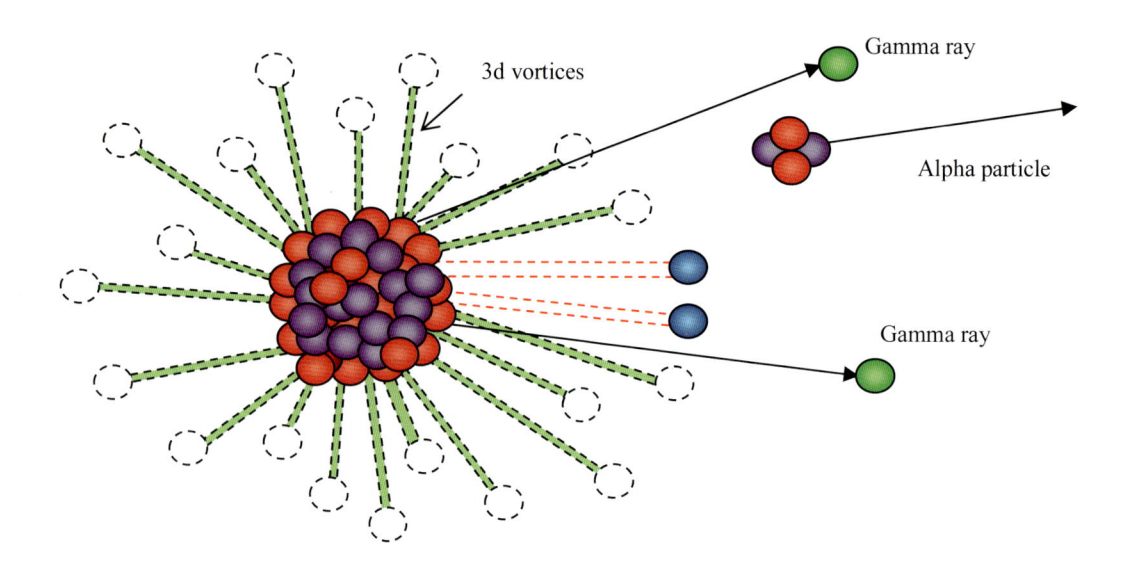

Part 4 The creation of the X-ray

If the electrons were closer to the nucleus, their vortices are shorter; hence when the Alpha particle is expelled, and the two vortices continue to flow into the nucleus, then they flow out; and being of lesser volume, become the x-rays seen in Figure 23.5 below…

Figure 23.5

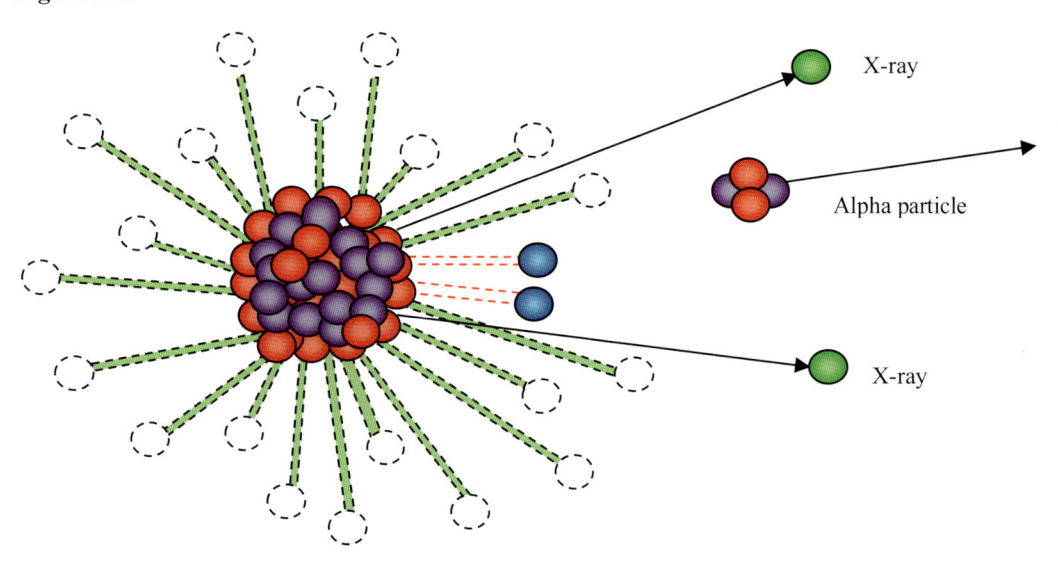

Most of what has just been said is shown in the diagram in Figure 23.6 below. Notice also how the two electrons in the atom and the two protons in the alpha particle are still connected via their fourth dimensional vortices [thick dotted lines].

Figure 23.6

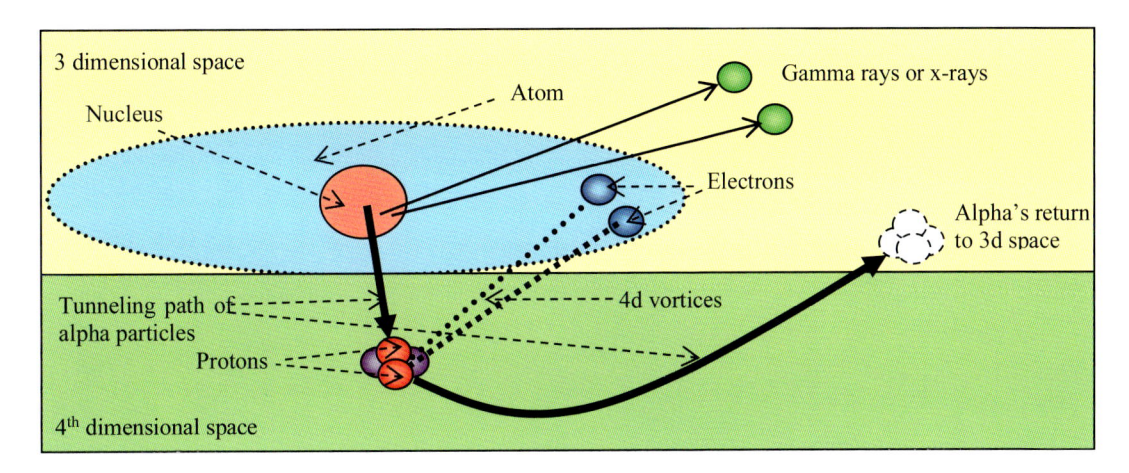

When the alpha particle returns to the three dimensional surface, its two protons are still connected to the electrons via their fourth dimensional vortices. However, they are no longer connected to the electrons via their three dimensional vortices. The volumes of three dimensional space within the three dimensional vortices that connected the electron to the proton continue on from the electron to the position in the nucleus where the two protons were; then out of the nucleus and into three dimensional space; becoming two gamma rays.

However, if the electrons were in shells closer to the nucleus, their vortices would be shorter and the volumes of the photons would be less; and instead of gamma rays being emitted from the nucleus, x-rays would be emitted. Either way, the freed unattached electrons that are left, give the atom a charge of – 2, [that is if they are not also thrown free]. The 3d space flowing back into the protons gives the alpha particle its + 2 charge.

Consequently, the mechanism via which the alpha particle and gamma rays [or x-rays] are emitted from the atom can be traced to the expansion and contraction of the vortices as the electrons move closer then away from the nucleus; causing an expansion of the volumes in their interconnected protons within the nucleus; causing an alpha particle to be "compressed" out of the nucleus.

It should also be noted that this "resonance" of the surrounding electrons happens rarely, causing the expulsion of the alpha particle to be a rare occurrence for one atom. However, because there are billions of trillions of uranium atoms in a few ounces, the R238, atoms are constantly decaying and emitting alpha particles with seemingly constant regularity.

Figure 23.7 [The final outcome…]

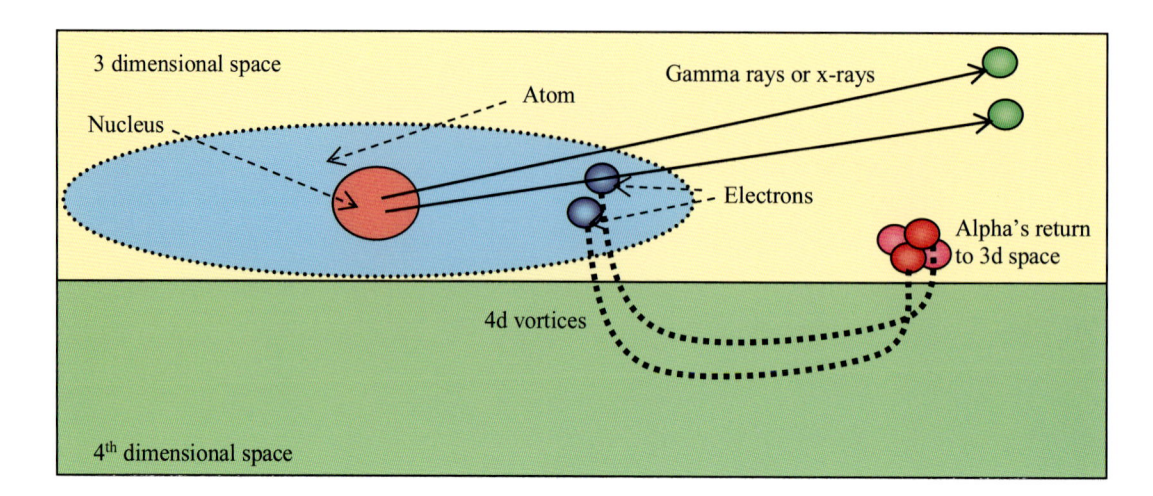

In the above Figure 23.7, notice that after the alpha particle returns to the three dimensional surface, its two protons are still connected to the two electrons via their fourth dimensional vortices. However, they are no longer connected in 3d space by 3d vortices; hence their "charges" return as 3d space again begins to flow back into the protons and out of the electrons. [Note: the vortices are not this curved shape but only drawn this way to show they go into and out of 4d space.]

PART VI
STRANGE RELATIONSHIPS PREVIOUSLY EXPLAINED ONLY BY MATH!

There are many strange relationships in science that previously could only be explained by mathematics. These are some of the more famous ones that can now be illustrated, such as Planck's Constant…

Chapter 24
Planck's Constant

The great German Physicist's constant can now be explained by examining how the volume of the space within the photon and its frequency of vibration are related.

In the formula $E = hv$, h = Planks constant, E = energy, and v = frequency of light. This formula can also be expressed as $h = E / v$; revealing that Planck's constant is directly proportional to energy and inversely proportional to frequency. Since h is a constant, if the frequency of a photon increases, its "energy" has to increase; and if the energy decreases, the frequency has to decrease.

All of this mathematics can now be explained with illustrations!

It is easy to see that the greater the volume of three dimensional space the photon has, *the greater and quicker will be its expansion and subsequent contraction.* This will make its frequency increase, giving Planck's Constant a simple yet eloquent explanation.

Figure 24.1

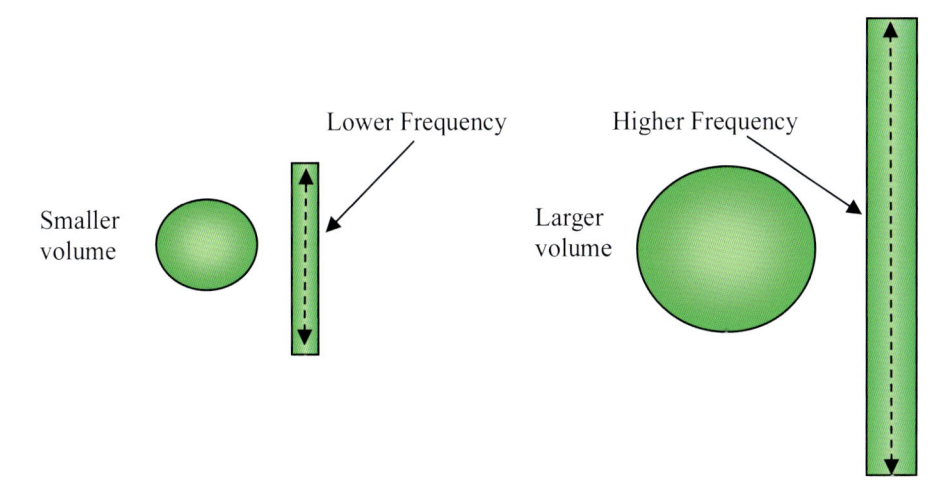

Because the vortex theory reveals that energy, or rather photons are nothing more than "packets" of very dense space surrounded by a massive region of dense space, created by the reconfiguring space it is passing through, Planck's constant is revealed to be the ratio between the amount of dense space in the photon we call energy and its quickness of expansion and contraction: its vibration.

When the photon expands and elongates, it becomes *less* dense. As it does, the surrounding space that was pushed outward by the presence of the photon begins to move inward, towards the photon. As it does, it becomes less dense, decreasing the intensity of the magnetic field's flux. When the space directly above and below the expanded photon begin to push against the now less dense space inside the elongated photon, it begins to contract. The space within the photon becomes denser, pushing outward upon the surrounding space, increasing its density – increasing the intensity of the magnetic field's revolving flux, expanding its size, increasing the intensity of its magnetic field.

> There are many strange relationships in science that previously could only be explained by mathematics. Here is the illustration of Einstein's famous equation!

Chapter 25
Illustrating Einstein's Famous Equation [E = mc²]

Einstein's famous equation illustrated?! It does not seem possible! But nevertheless it is true. In fact the illustration is so simple it is almost laughable! It works like this…

Take a fresh tub of ice cream and a round scoop. Then carefully remove a scoop from the top of the ice cream. The hole in the ice cream represents matter; the ice cream in the scoop represents energy! It is a simple as that!

Figure 25.1

Tub of ice cream with scoop taken out of it: the scoop of ice cream!

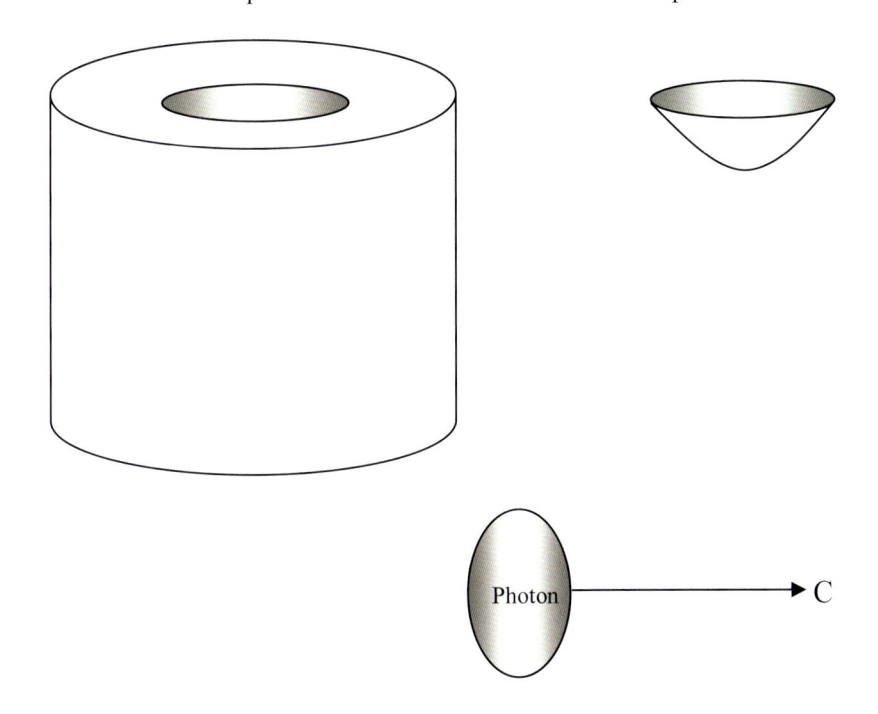

The *hole* in the ice cream represents the holes that matter is created out of. The *scoop* of ice cream represents the packet of dense 3d space photons of energy are created out of! The speed of light "C" represents the speed of photons of energy.

> There is no particle responsible for the phenomenon of mass. The phenomenon of mass is created by the surface tension of three and higher dimensional holes resisting an outside force trying to distort them.

Chapter 26
Mass

Mass is not created by a particle! There is no Higgs Boson particle. The true explanation of mass begins and ends with distortions of the surfaces of the three dimensional, and higher dimensional holes we call matter.

Below is an enlarged cross-sectional drawing of an electron and a positron. Notice how the shape of a sphere is maintained by the inside pressure (blue) equaling the outside pressure (red). Because these pressures are equal all over the surface of the holes, they maintain a steady surface tension within and without. Note: $\overline{p}$ = positron.

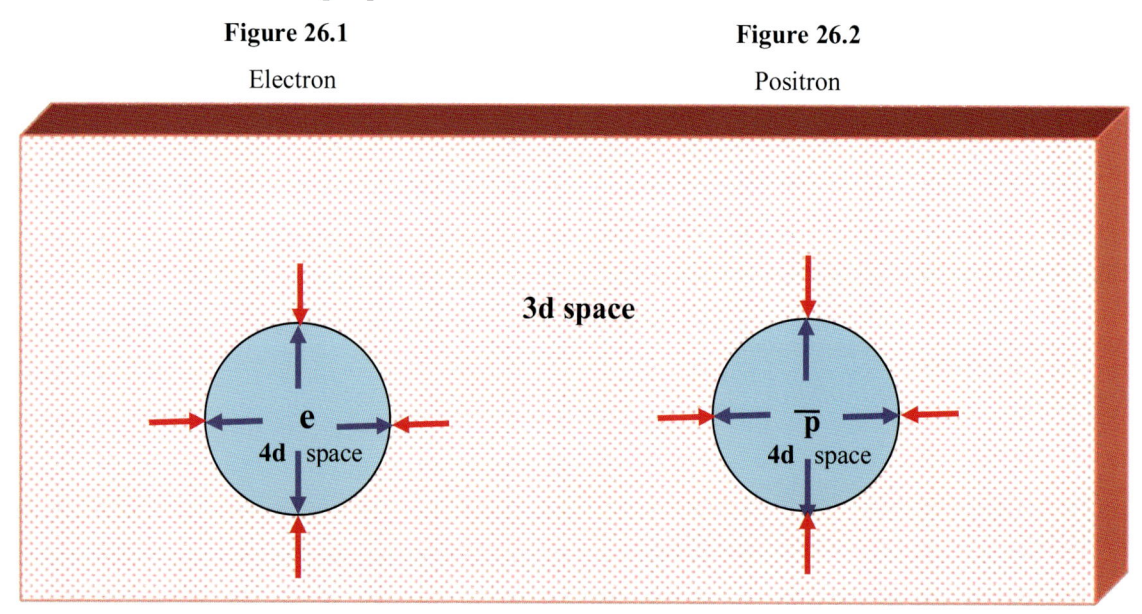

Figure 26.1 **Figure 26.2**
Electron Positron

It must also be realized that the spherical shape is a function of the *surface tension* of the inside 4d space. The harder it is to distort its surface creates the phenomenon known as MASS. In the Figure below, the surface of the positron is distorted…

It does not take too much of a force to distort a positron. However, if a positron contains quarks turning it into a proton, a force not only has to distort the surface of the positron but the surfaces of the 5d holes within it that are the quarks…

A **proton is an inflated positron** containing three quarks: two Up quarks and one Down quark. These quarks are 4d holes existing upon the surface of the 4d volume of space within the interior of the positron. Note how these 4d holes contain within them volumes of 5d space [pink]. The proton's +1 charge comes from the positron and not the quarks. Quantum-chromo dynamics is a mistake!

Figure 26.3

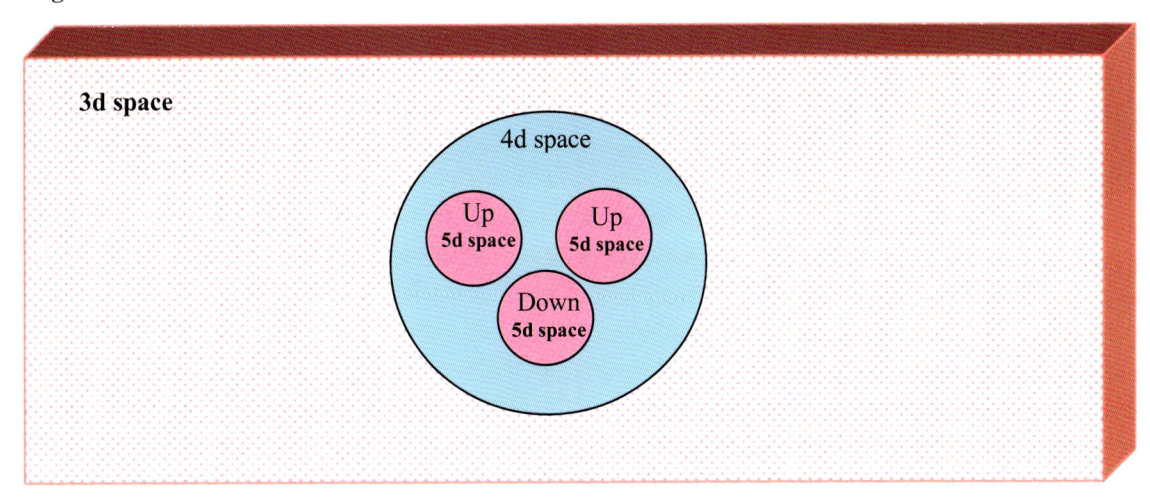

The reason why the proton is much more "massive" than the positron is that to move it, not only does the surface of the proton need to be distorted, but also, all of the surfaces of the quarks in the interior. This is much more difficult to do because the 3d space outside the proton has to distort the 4d space inside the proton, that in turn has to then distort the surfaces of the quark's 4d holes.

Figure 26.4

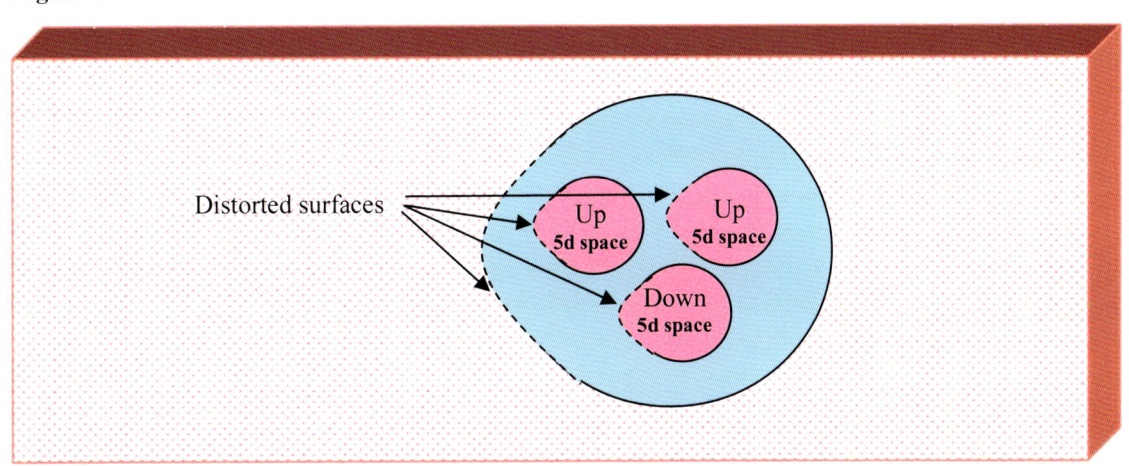

Because of the hierarchy of holes, a Charm quark is a fifth dimensional hole within an Up quark; while a Strange quark is a fifth dimensional hole within a Down quark. Hence to move such a particle containing fifth dimensional holes within six dimensional space is even harder, making such a particle seem even more "massive".

Figure 26.5   **Figure 26.6**

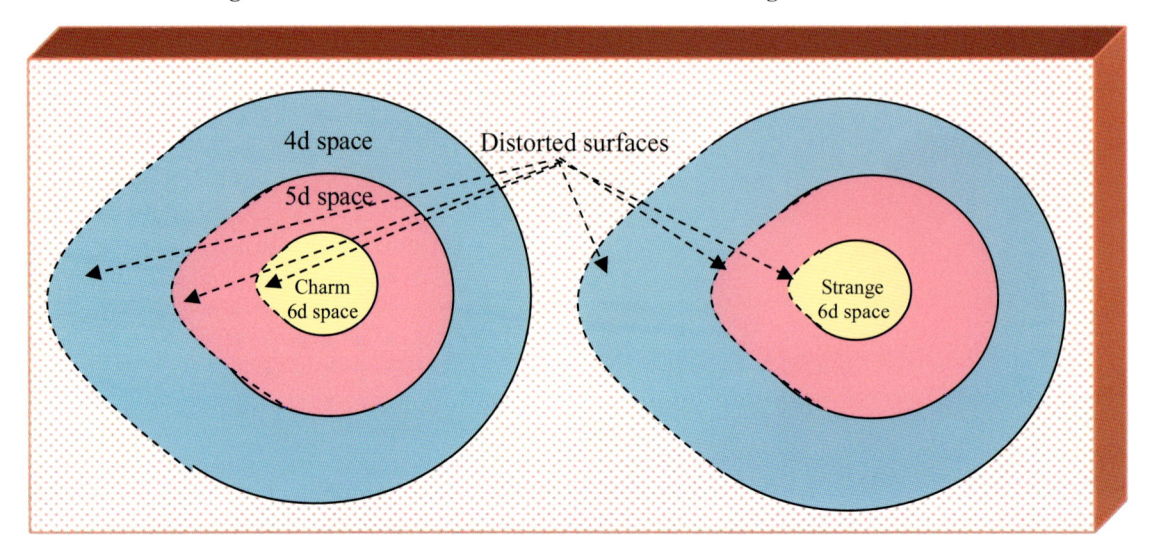

To move a particle containing a 6d hole such as a **Bottom Quark** within 7d space is super hard, making such a particle very "massive".

Figure 26.7

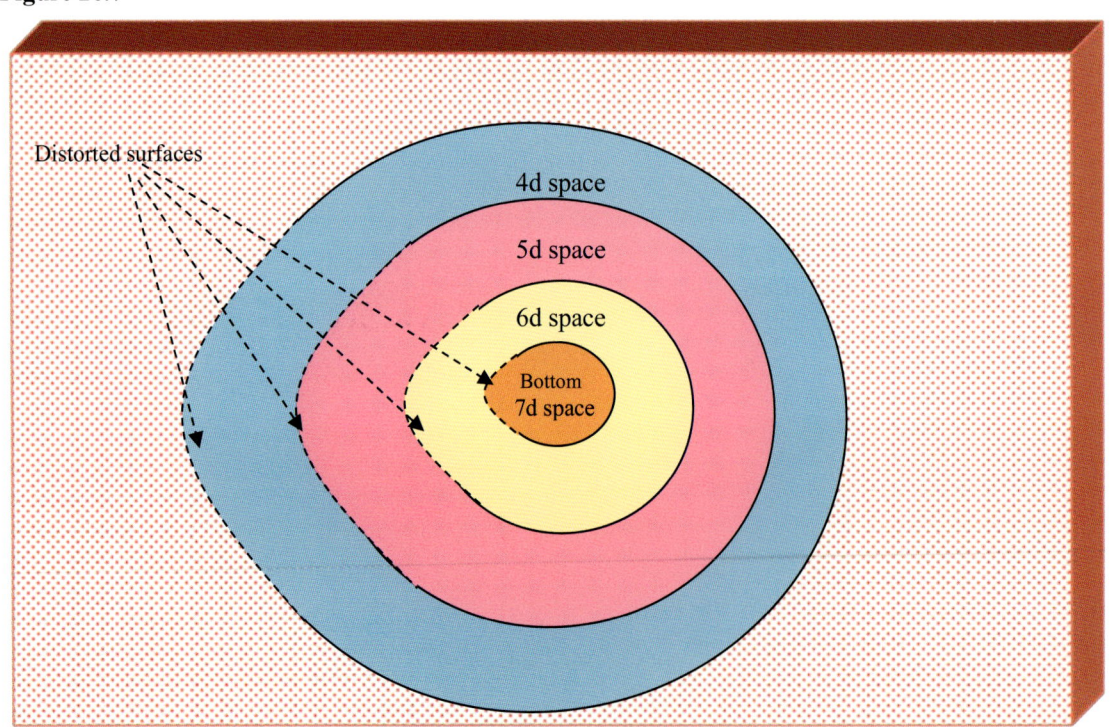

To move a particle containing a 6d hole such as a **Top Quark** within 7d space is super hard also, making such a particle seem to be the most "massive" of all.

Figure 26.8

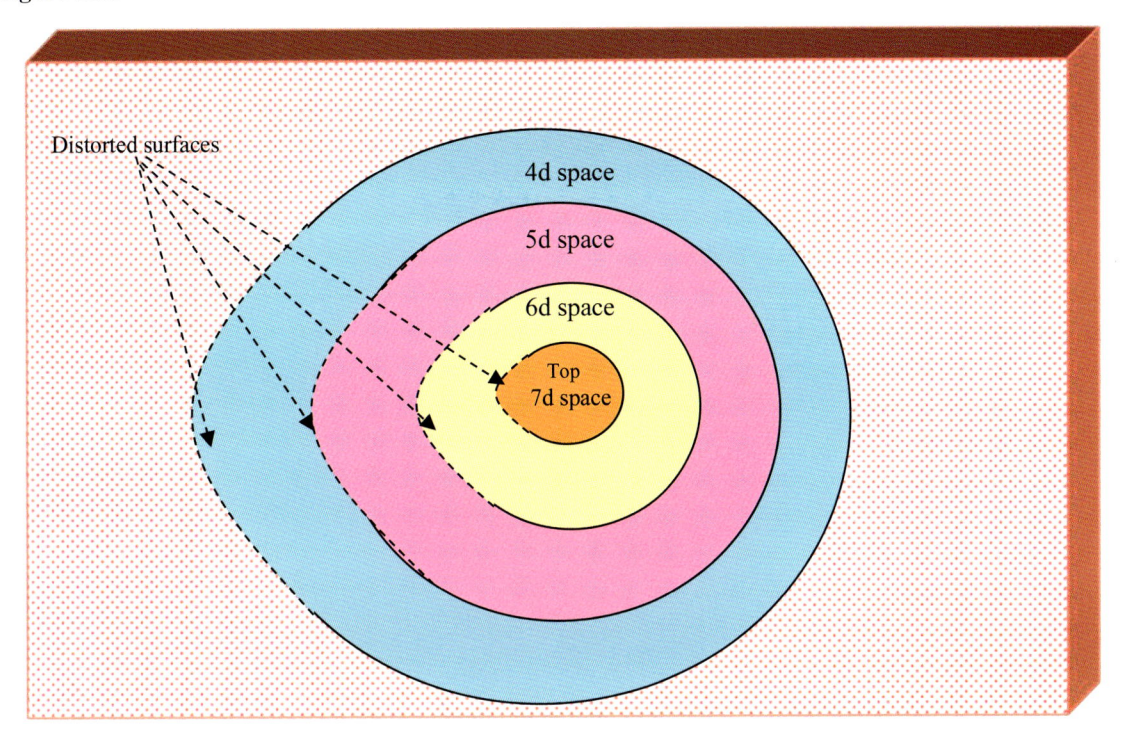

Hence, it is now easy to see that it is the surfaces of "particles" of matter resisting distortion that make them harder to move, creating the phenomenon of mass.

> Ever since Newton developed his three laws of motion 300 years ago, they were acknowledged to be axioms: unprovable observations. But with the coming of the Vortex Theory of Atomic Particles all of that has changed. Gravity is explained too.

Chapter 27
Newton's Law of Gravity: $F = G\, m_1 m_2 / r^2$ And His 3 Laws of Motion & $F = ma$

Shockingly, Newton's *law of gravity* has never been understood! Everyone considers it to be a "pulling" force; a force that "pulls" two objects together: that the apple is pulled by gravity towards the surface of the earth. But this is a mistake! The shocking reality is that it is really a "pushing" force that pushes two objects together! It happens like this…

GRAVITY AGAIN …

Let us imagine that two objects such as two stars are attracted to each other by their mutual gravity. The twentieth century conventional science's explanation envisions them being pulled together. In the following drawing notice how Star #1 pulls on Star #2; and Star #2 pulls on Star #1.

Figure 27.1

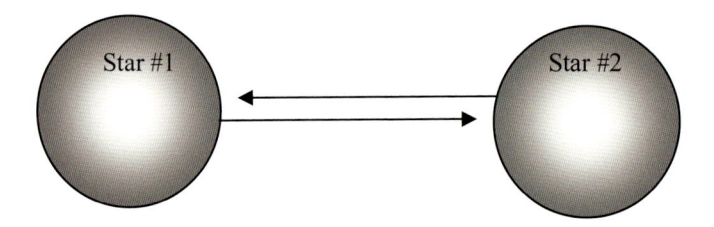

BUT THIS IS NOT WHAT IS REALLY HAPPENING!!!

To understand what is really happening, let us look at an individual proton within each star. Furthermore, let us envision the density of space as being represented by lines. Then in the drawing below, notice how space is less dense between the two protons and dense on the opposite sides…

Figure 27.2

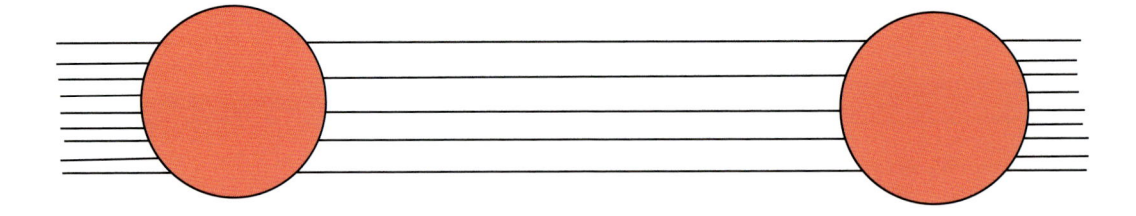

The less dense space causes the sides of the two protons facing each other to distort outwards towards each other: [note, the distortion is greatly exaggerated].

Figure 27.3

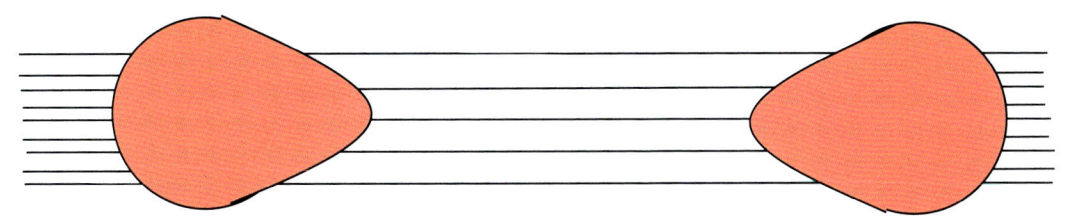

But this distortion creates a problem. Because of the surface tension, the 3d hole tries to straighten itself out. This causes the back side to move towards the distorted side:

Figure 27.4

However, the instant they try to re-establish their spherical shape the sides facing each other are distorted again, causing the two protons to be slightly accelerated as they move closer to each other.

Figure 27.5

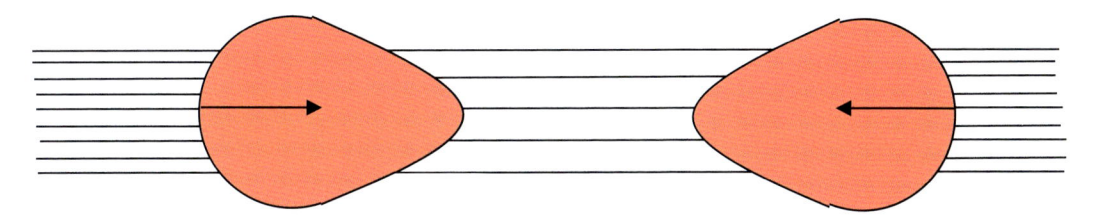

And now it can be seen that gravity is not "pulling" two objects towards each other, but rather, causes them to "push" themselves towards each other!

Because the strength of any "field" dissipates with the square of the distance this accounts for [r^2] in his equation. The gravitational constant has to take into account many factors. But the fact that gravity is so weak in comparison to the other forces of nature, and there is nothing to equate it to, it has to be determined by experiment only; first done by Henry Cavendish in 1798 using his torsion balance apparatus.

NEWTON'S FIRST LAW

There are slight variations in describing Newton's first law; the following version appears to be one universally accepted:

An object at rest stays at rest and an object in motion stays in motion unless acted upon by an unbalanced force.

The first half of this law is easy, an object at rest stays at rest because the pressures without and within are balanced [Figure 27.6]. The second half of the first law, stating that an object in motion stays in motion unless acted upon by an unbalanced force, is easily explained by waves created by space reconfiguring itself both inside and outside the sub-atomic particles an object is made out of [Figure 27.7]. As said above, as the front of the hole is distorted outward, the back side tries to straighten out the hole by moving forward. But as soon as it does, the external wave created in the 3d space in front of the hole immediately distorts it outward again; and the sequence continues to repeat unless a force or object hits the hole and disrupts the two waves.

[Note: it has to be stated that nothing is at rest in the universe, everything is in motion. However for the sake of simplicity, the following illustrations are all hypothetical.]

Figure 27.6 **Figure 27.7**

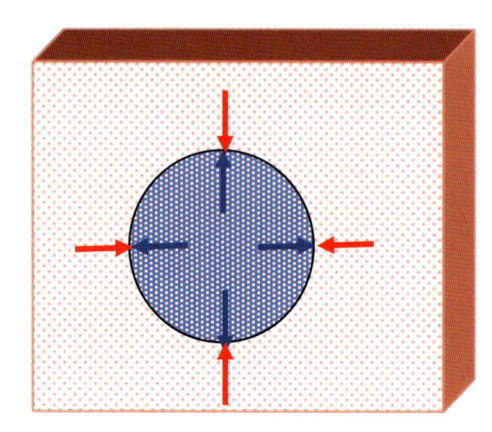

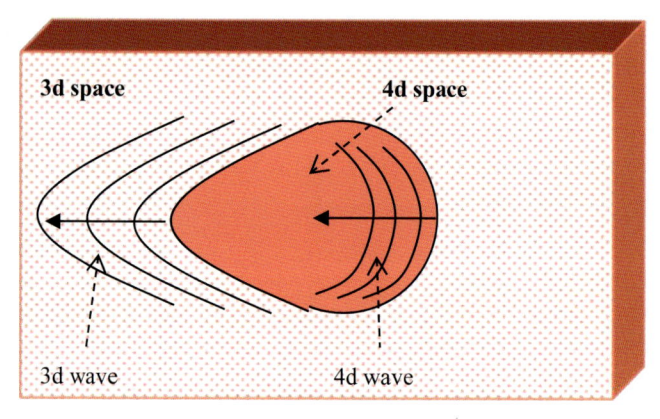

NEWTON'S SECOND LAW

Newton's second law is usually expressed thusly: the force F on an object is equal to the mass m of that object multiplied by the acceleration of the object: $F = ma$.

This law is easily explained by understanding the mass [m] of the object represents the number of sub-atomic particles an object possesses. While the acceleration [a] of the object represents the amplitude of the distortion the particles possess. This can be illustrated in Figure 27.8 below. The amplitude of the distortion equals the amount of acceleration a particle possesses. The larger the amplitude, the larger the acceleration.

Figure 27.8

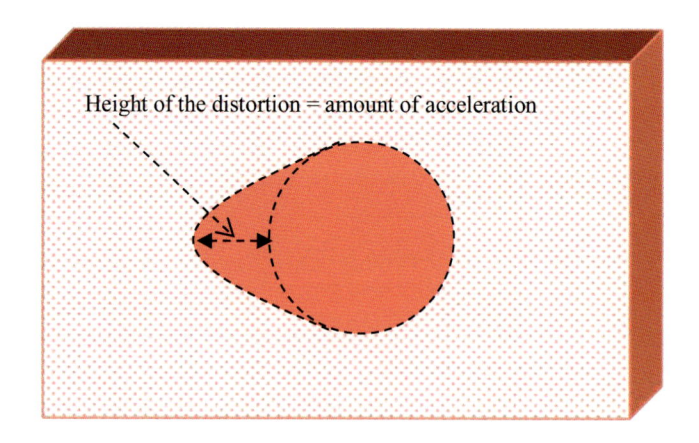

NEWTON'S SECOND LAW

Note: the distortions in the holes of one object are transferred to the holes of the second object. If the object being hit has less holes than the object hitting it, the more distortions are transferred, and the hit object moves quickly; but if the hit object has more holes than the hitting object it moves slowly. The other variable is the velocity of the hit. The faster the motion of the hitting object, the larger the distortions in the hitting object and a greater amount of distortions are transferred to the hit object and the faster it moves.

In a perfect elastic collision, the Figure below illustrates how the momentum of one object with its set of molecules, striking another object with an identical set of molecules can be explained with the transfer of the distortions in the molecules of the striking object to the struck object:

Figure 27.9

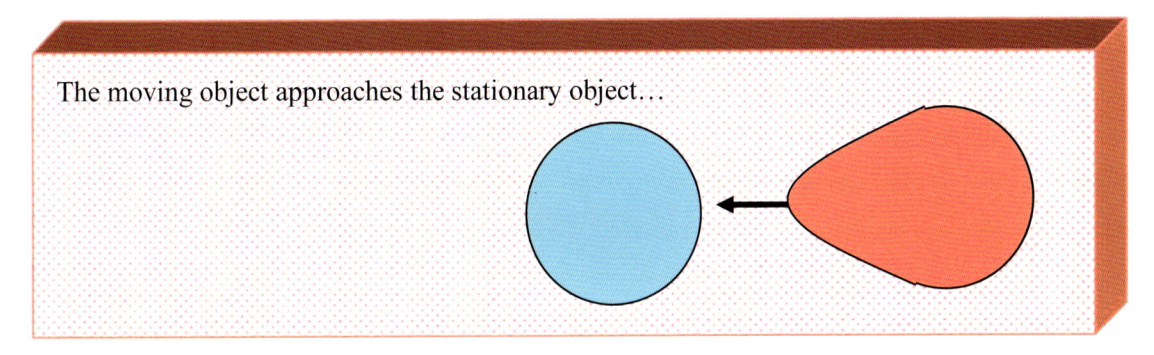

The moving object approaches the stationary object…

Figure 27.10

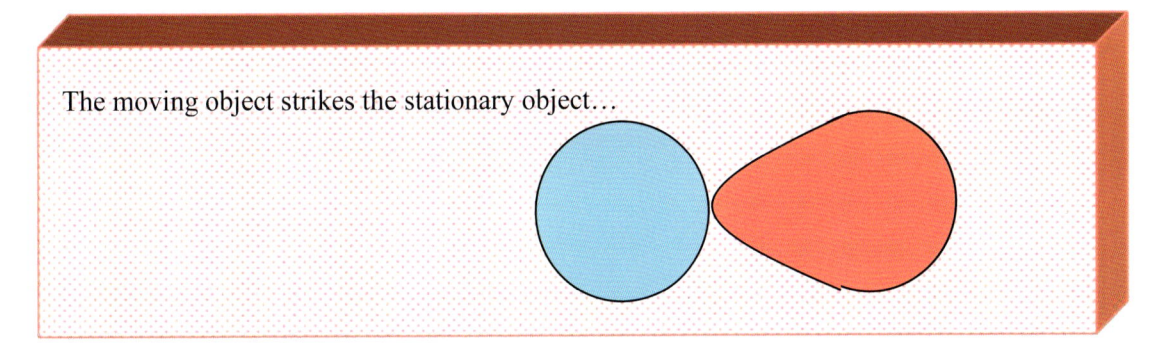

The moving object strikes the stationary object…

Figure 27.11

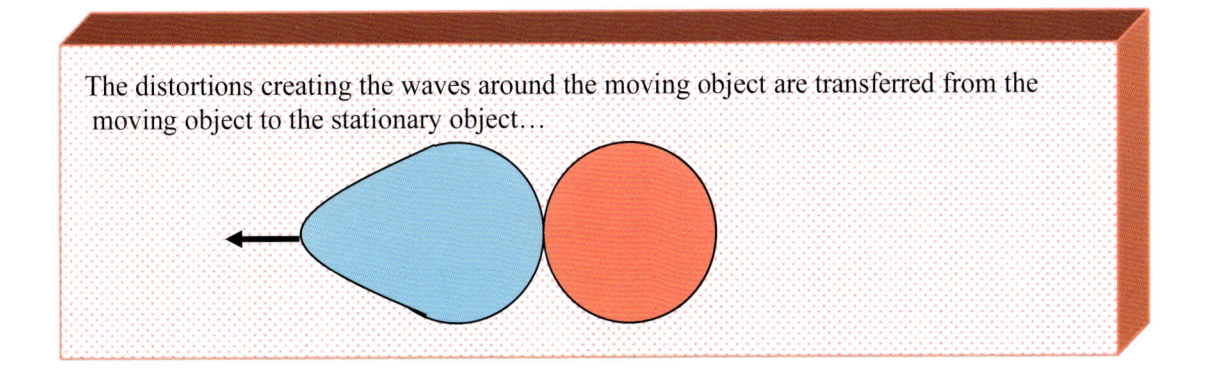

The distortions creating the waves around the moving object are transferred from the moving object to the stationary object…

Figure 27.12

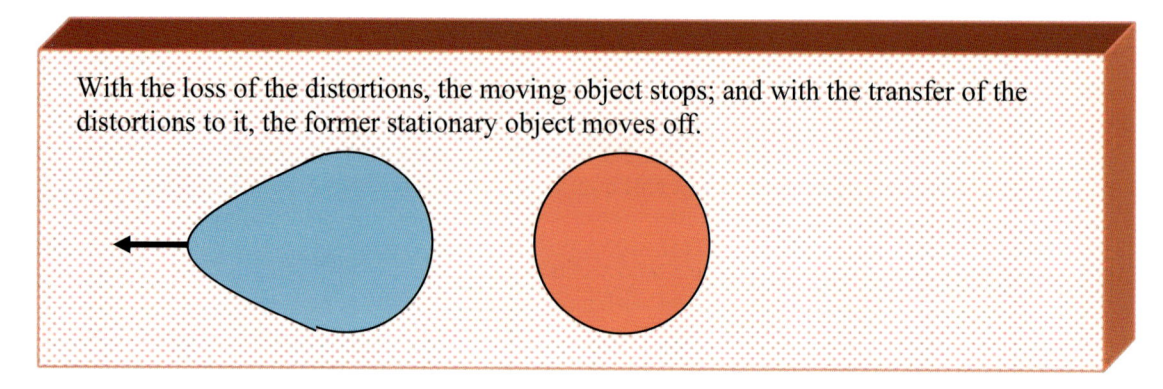

With the loss of the distortions, the moving object stops; and with the transfer of the distortions to it, the former stationary object moves off.

ANGULAR MOMENTUM…

Angular momentum is explained in Book 2. It is found in the illustration called the ball and the string. It results in the distortions drawn below.

The angular momentum can be explained as the vector addition of two different distortions that add up to create one distortion.

A ball whirling on a string creates two distortions. The addition of these two distortions can be seen in the following drawings: Figure 27.13 shows the distortion if the hole was moving through space in a straight line; Figure 27.14 shows the added distortion created by the pull from the string; Figure 27.15 shows the vector addition of the two distortions; and Figure 27.16 shows the resultant vector distortion.

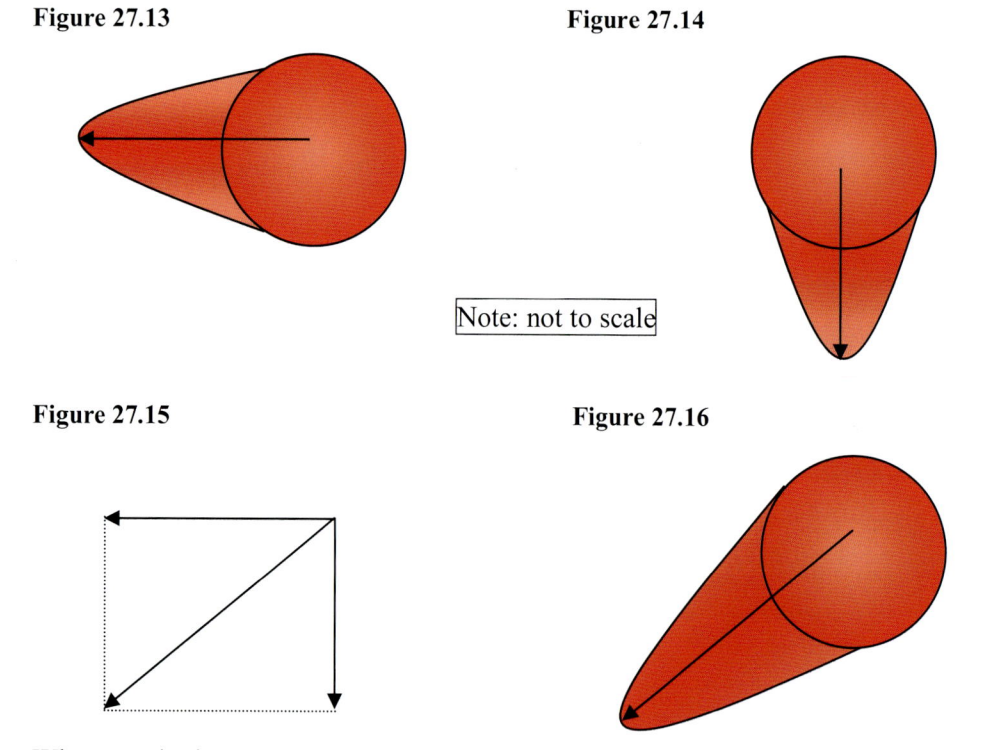

Figure 27.13   **Figure 27.14**

Note: not to scale

Figure 27.15   **Figure 27.16**

When one body exerts a force on a second body, the second body simultaneously exerts a force equal in magnitude and opposite in direction on the first body.

THE THIRD LAW: FOR EVERY ACTION, THERE IS AN EQUAL AND OPPOSITE REACTION.

Newton's third law of motion is explained using the above explanation for the holes out of which matter is constructed in Chapter 26 resisting distortion. For example, if an astronaut in space pushes against the space station. Since it is constructed out of thousands of more holes than the man, the distortion of the holes in his body are transferred to the holes in the space station. However, because the ratio of the amount of holes the station is constructed out of to the holes in the man's body are so large, they are instantly reflected back to the holes the man is made out of and he is accelerated backward.

However too, if two astronauts of equal weight have an equal number of holes both men are constructed out of, a different situation occurs. When one man pushes against the other man, the distortions of the holes his body is made out of tries to import to a distortion to the holes the other man is made out, their internal surface tension and the external pressure of the outward space upon them resist the distortion. Hence, most of the bend is instantly reflected back to other man. But then, the surface tension in the holes his body is made out of and the external pressure of the surrounding space pressing upon them instantly causes most of the bend to be reflected back to the opposite man. And on and on it goes in an instant of time until the half of the distortion that originally tried to accelerate the holes the second man is made out of are transferred. Consequently, both men are propelled backward at equal velocities.

> Quantum mechanics is acknowledged to be one of the most difficult subjects in all of science to comprehend; especially Schrödinger's equation. However, in actuality, its explanation is perhaps one of the easiest!

Chapter 28
Schrödinger's Equation: And the Foundation for All of Quantum Mechanics

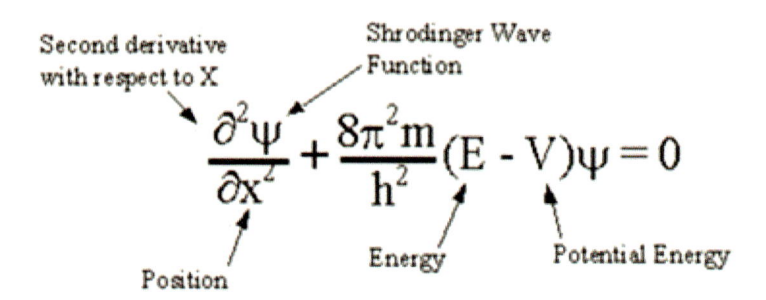

$$\frac{\partial^2 \psi}{\partial x^2} + \frac{8\pi^2 m}{h^2}(E - V)\psi = 0$$

- Second derivative with respect to X: $\frac{\partial^2 \psi}{\partial x^2}$
- Shrodinger Wave Function: ψ
- Position: x
- Energy: E
- Potential Energy: V

WOW! What an equation! Just looking at this conglomeration of figures is intimidating to a first semester student of physics. But what is even more shocking is that according to the Vortex Theory of Atomic Particles what it is describing is a phenomenon that is simply and easily explained!

PART I

THE WAVE ASPECTS OF SPACE EXPLAIN THE FUNDAMENTALS OF QUANTUM MECHANICS

Returning our attention to the fable of the blind men and the elephant, we recall that the blind men each touched a different part of the elephant and described it as something different from what everyone else was describing. The same problem exists in the case of matter.

One scientist does an experiment and says… the electron is like a particle; while another scientist does another type of experiment and says the electron is like a wave.

The truth is…each is right, and each is wrong. Here is the answer…

THE ELECTRON IS LIKE A PARTICLE…

The electron has mass! This truism was discovered in 1897 by J.J. Thompson during experiments using cathode ray tubes. He called them "corpuscles", a name that was later changed to the "electron" by George Stoney

THE ELECTRON IS LIKE A WAVE…

In 1924, French scientist Louis de Broglie hypothesized that matter can be represented as a wave. This led to Erwin Schrödinger's 1926 famous wave equation, creating the foundation for quantum mechanics.

THE ELECTRON AS A PARTICLE…

Notice how the 3d hole that is the electron has space flowing out of it creating its electrostatic charge.

Figure 28.1

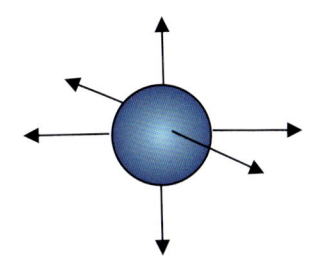

THE ELECTRON AS A WAVE…

Note in the figure below how a volume of denser 3d space surrounds the electron. This volume is created as the electron moves through 3d space and is rotating as the electron rotates.

Figure 28.2

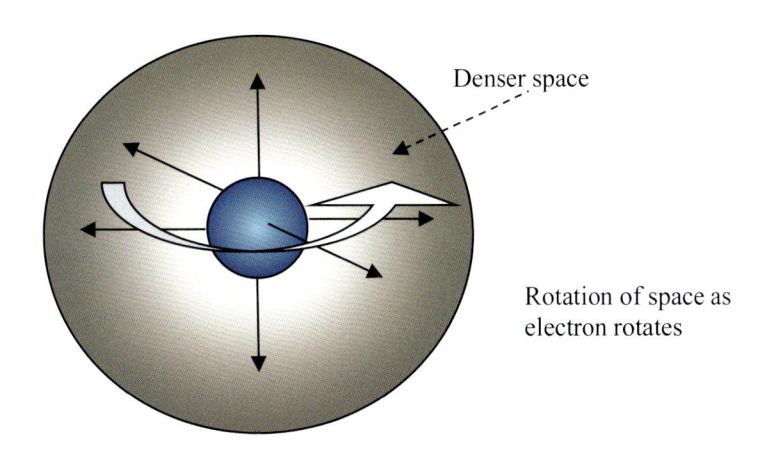

Denser space

Rotation of space as electron rotates

THE ELECTRON AS A WAVE: NOTE THE WAVE IS CREATED
IN THE <u>SURROUNDING</u> 3D SPACE…

In the Figure below, we see that as the electron moves through 3d space, the pressure of the outward flowing 3d space known as its electrostatic charge causes the surrounding space to expand outward as the electron approaches then contract back inward as the electron passes. Creating a wave like pattern…

Figure 28.3

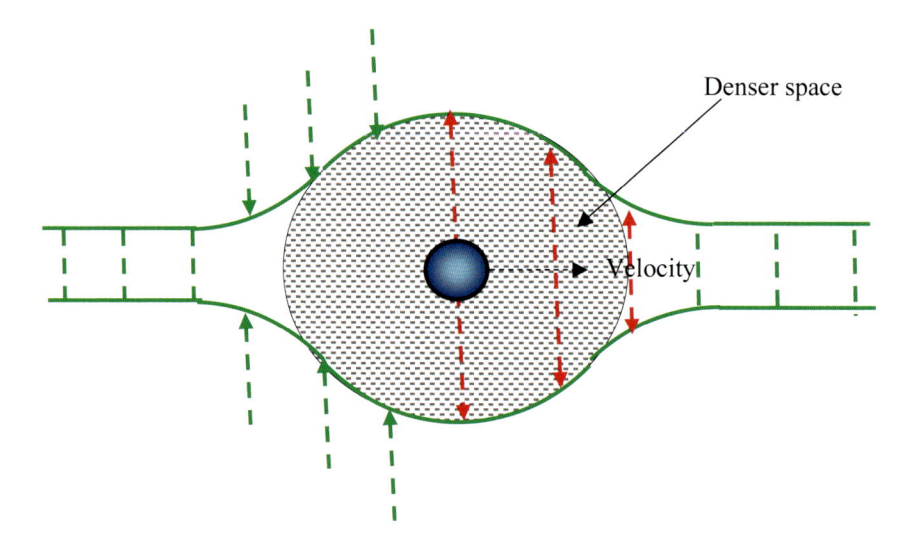

THE PROTON AS A PARTICLE

The hole that is the proton has space flowing into it creating its electrostatic charge.

Figure 28.4

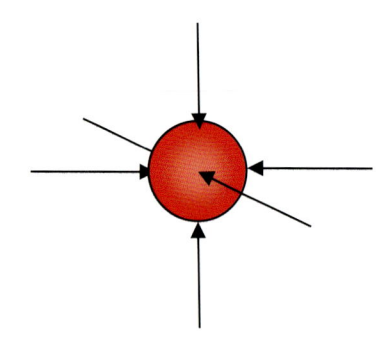

THE SURROUNDING 3D SPACE IS PULLED INWARD, BECOMING LESS DENSE…

Note in the figure below how a volume of less dense 3d space surrounds the proton. This volume is created from the surrounding space as the proton moves through 3d space.

Figure 28.5

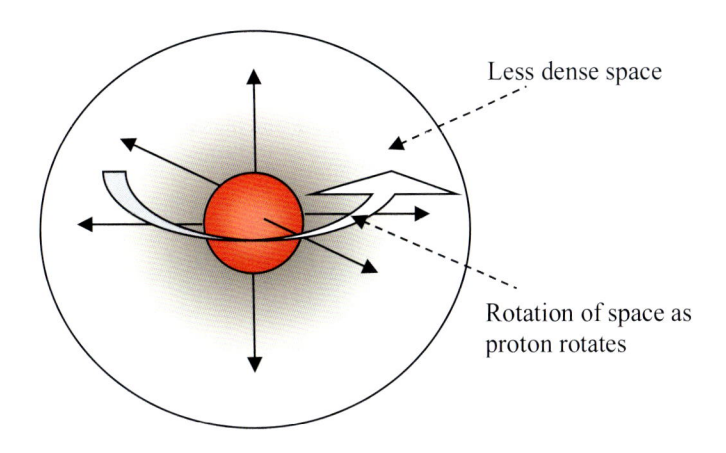

THE PROTON AS A WAVE: NOTE THE WAVE IS CREATED AS THE PROTON PULLS
<u>THE SURROUNDING</u> 3D SPACE INWARD…

For the proton the opposite is true. Because the proton pulls 3d space inward, it creates a region of less dense space surrounding it. This less dense region of space is created when the surrounding 3d space is first pulled towards the proton, then contracts back as the proton passes…

Figure 28.6

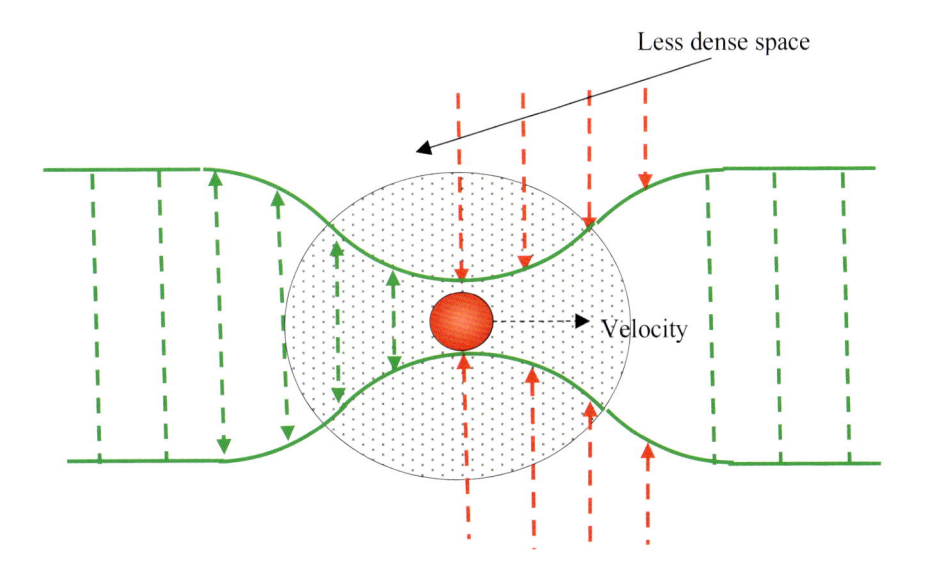

PART II

THE PARTICLE AND WAVE ASPECTS OF ENERGY ALSO EXPLAIN THE FUNDAMENTALS OF QUANTUM MECHANICS

Again returning our attention to the fable of the blind men and the elephant, we see that the same problem that developed with matter happened with energy too. One scientist does an experiment and says… the photon is like a particle; while another scientist does another type of experiment and says the photon is like a wave.

THE PHOTON IS LIKE A PARTICLE AND A WAVE!!!

Albert Einstein discovered the photoelectric effect in 1905. Because of this experiment it was then concluded that photons behave like tiny "packets" of energy.

Years before in 1801, Thomas Young performed his famous "Two Slit Experiment" that showed light behaved like a wave. And again, just like the fable of the blind men and the elephant, both observations are correct!

THE PHOTON AS A PARTICLE…

Notice how the condensed 3d space the photon is constructed out of makes it seem like a particle.

Figure 28.7

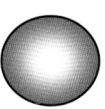

THE SURROUNDING 3D SPACE IS PUSHED OUTWARD, BECOMING DENSER…

But again, as we have seen earlier in Chapter 19-20, the photon as a condensed region of space that pushes the surrounding space outward as it moves. In the Figure below, note how a volume of denser 3d space surrounds the photon. This volume is created as the photon moves through 3d space and the surrounding 3d space *reconfigures* around it.

Figure 28.8

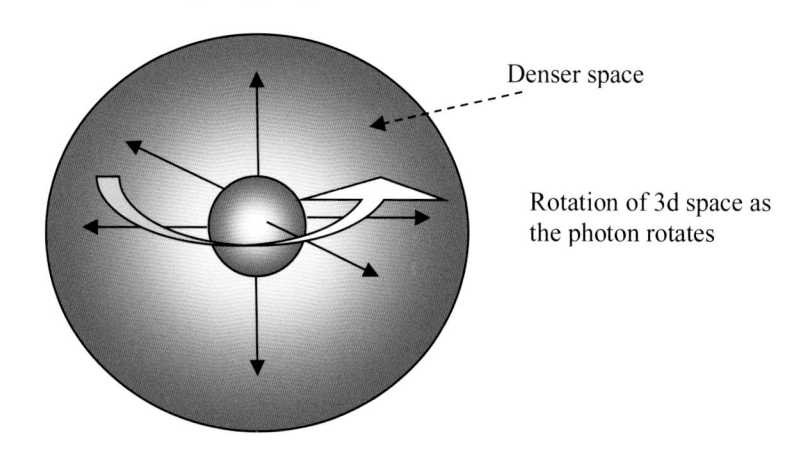

Denser space

Rotation of 3d space as the photon rotates

The photon as a wave: note the wave is created
in the <u>surrounding</u> 3d space…

It is this reconfiguring of space that the Schrödinger equation is based upon.

PART VII
BIZARRE PEHONMENON EXPLAINED; AND THE EXPOSURE OF FALSE IDEAS

Young's two slit experiment explaining the wavelength of light was perhaps one of the greatest experiments ever performed. It showed that light behaves similar to a wave in water. However, this great experiment has never been understood until now. It was not a wave that the photons of light possessed, but rather waves created in the surrounding space that creates the effects!

Chapter 29
The Explanation of the Double Slit Interference Patterns Created by Single Electrons and Photons!

The mystery of how one electron appears to go through the two slits at the same time and creates an interference pattern is explained by the large dense region of space that surrounds protons and electrons. This denser region of space is created by the reconfiguration of space behind the two slits as the photon or electron approaches them. this reconfiguration process proceeds the electron or photon through the two slits, creating alternating corridors of denser and less denser space for the electron or photon to proceed along.

The great mystery of how a single electron can create the wave like interference pattern in Thomas Young's famous twin slit experiment is easily explained by the Vortex Theory of Atomic Particles.

The wave-like interference pattern made by projecting either light or matter through the twin slits *is created by the dense regions of three dimensional space reconfiguring around photons or electrons.* In comparison to the size of the electron or photon, the dense region of space that surrounds it is massive.

Figure 29.1

Note: in relation to the size of the electron, the region of less dense space that surrounds an electron is so massive it is impossible to draw the proportionate sizes.

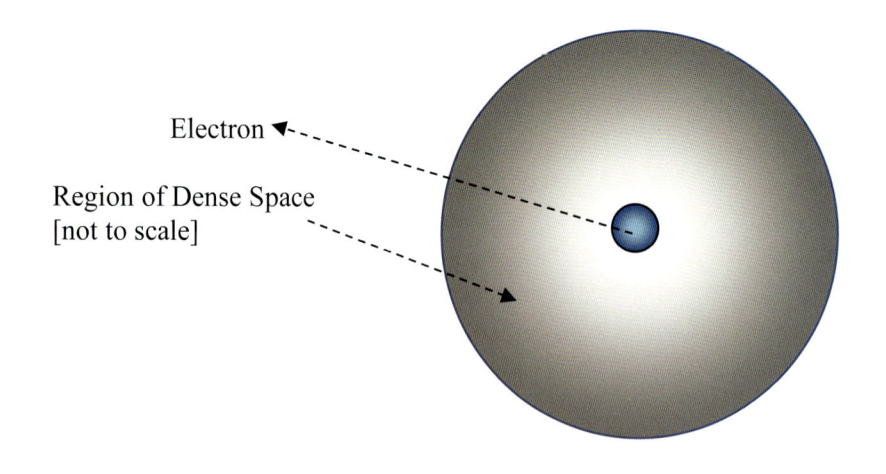

This region is so massive, that when an electron approaches the twin slits, this region of *reconfiguring space* arrives first.

Figure 29.2

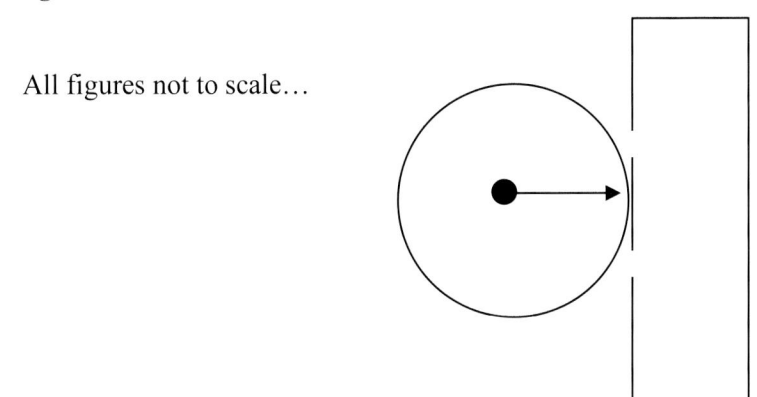

All figures not to scale…

Because the dense region of reconfiguring 3d space arrives first, it causes the 3d space directly behind the two slits to begin reconfiguring.

Figure 29.3

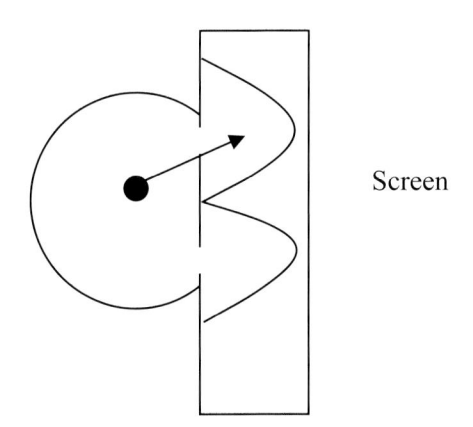

Screen

As this denser region of reconfiguring 3d space moves through both slits simultaneously, the two waves interfere with each other and begin to create the invisible wave patterns of dense and less dense space on the screen.

Figure 29.4

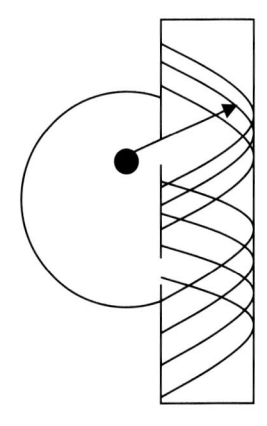

When the electron finally moves through one of the slits, it moves into a region of space that is alternately dense, less dense, dense, less dense, etc. Consequently, the surface of the electron is distorted towards *one of the less dense regions,* avoids the denser regions, and travels in the direction of one of the "corridors" of this less dense region, striking the screen.

Figure 29.5   **Figure 29.6** 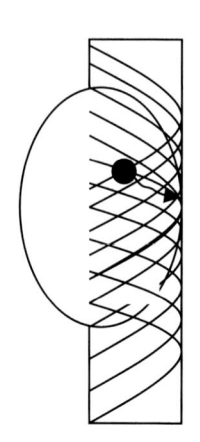   **Figure 29.7**

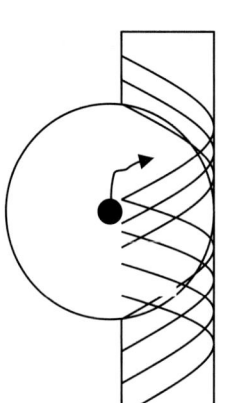

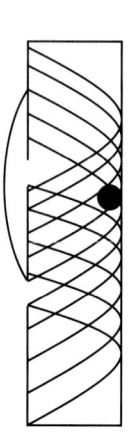

Note, in Figure 29.6, as the electron passes though one of the twin slits, note how the dense region behind the slit has not yet passed all of the way through. This follow through of the sphere of dense space behind the electron continues the interference pattern until the electron strikes the screen. Note too, this picture shows the electron going through the top slit. If the region of dense space were oriented more towards the bottom slit, it would have gone through that one.

When this same situation happens over and over again, it is their slight differences in their direction of travel that sends the electrons to the less dense regions, avoiding the denser regions, and creates the light and dark patterns seen on the screen.

Figure 29.8

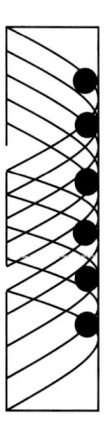

Because it can now be seen how a single electron can create the interference pattern, some important misunderstandings can now be laid to rest.

#1. Perhaps the most important misunderstanding is the mistaken idea that somehow the electron was thought to go through both slits at the same time. As can now be seen, this is not what is happening. The particle characteristic of the electron, [its 3d hole], will only allow it to pass through one hole.

#2. Equally important is the idea that photons of light are somehow interfering with each other, "canceling each other out" creating the dark spots on the screen. This is also a mistaken idea. Instead, it can now be realized that photons like electrons will avoid the denser regions and are attracted to the less dense regions.

> Dark matter, the great mystery of astronomy is easily explained. As a galaxy, or a cluster of galaxies rotate, the space that the sub-atomic matter is imbedded in is rotating too. Ending one of the great debates in 20th Century Astronomy.

Chapter 30
Dark Matter

According to Newtonian physics, the rotation of interstellar matter should be defined by the laws of gravity. For example, in our solar system, the inner planets such as Mercury rotate faster while those further out such as Jupiter rotate slower. This seems perfectly logical.

However, in the 1930's Dutch astronomer Jan Oort observing Doppler shifts of light in neighboring galaxies observed that stars on the outer edges of galaxies were moving too fast for conventional gravity to explain. Since the galaxies were not flying apart he reasoned that there had to be additional matter he could not see that was holding them together; causing the stars to rotate "un mass", more like a record on a turntable than as individual objects. Hence, the idea of "missing mass", later known as "dark matter" came into being.

Enter Fritz Zwicky of Caltech. He discovered the presence of dark matter on a much grander scale while using Doppler shifts of light to investigate the galactic cluster known as the Coma cluster, 300 million light years away. This massive cluster of galaxies first thought to be made up of 1000 galaxies, now appears to contain as many as 10,000! However, the amount of matter he calculated to exist within these galaxies was not enough to explain the force holding them together. He concluded that there had to be the presence of a tremendously large amount of missing matter he could not see. Hence, Dark matter!

Then, on and on it went. More investigations by other astronomers kept discovering more and more about Dark matter. But again, as explained before in this book, this is flawed logic. When Einstein made his false proposal that space was made of nothing, [and grown adults believed what he said next: that gravity was made of bent space: (and again, how do you bend something that does not exist?)], there was nothing left for rational scientists to do but reason illogically! They were forced to work with the mistaken premise that everything in the universe has to be made up of particles! Including the strange, and seemingly illogical phenomenon they were observing; and had dubbed it "Dark matter".

However, this misconception can now be cleared up with the discovery of the Vortex Theory of Atomic Particles. Using the basics of this theory that space is made of something, and proven via the *End of Time* thesis, it is now revealed that the phenomenon of "Dark matter" is being created by the rotation of space itself!!!

As a galaxy, or a cluster of galaxies rotate, the space that the sub-atomic matter is imbedded in is rotating too: like a gigantic intergalactic whirlpool! [For brilliant future scientists: like the story of the chicken and the egg, what came first? Was it a whirlpool in space that entrapped the matter; or was it the rotating matter that created the whirlpool?]

THE CREATION OF GALACTIC AND GALACTIC CLUSTER ROTATIONS…

In an Elliptical galaxy, the sub-atomic particles stars are made out of are embedded in a whirlpool of revolving space. The rotation of this space makes them, and the stars they create rotate like that of a record on a record player instead of all moving as separate individual objects held together only by gravity.

Figure 30.1

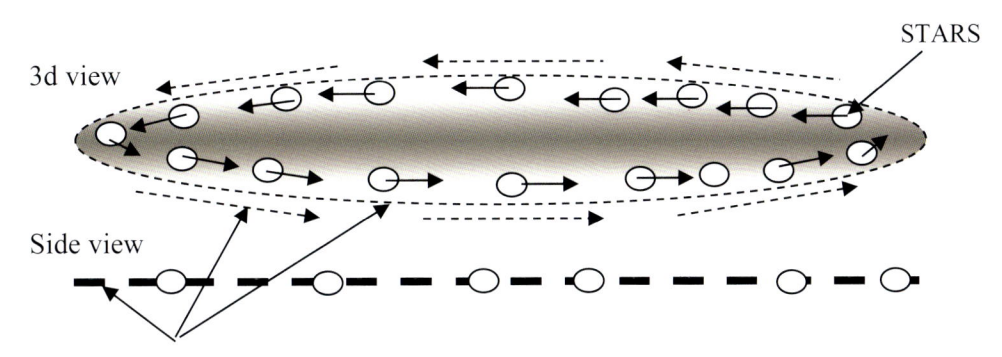

THE WHIRLPOOL OF SPACE

One picture equals a thousand words: in the picture below, the whirlpool of revolving space is what the sub-atomic particles the stars are made out of are trapped within and imbedded within…

Figure 30.2

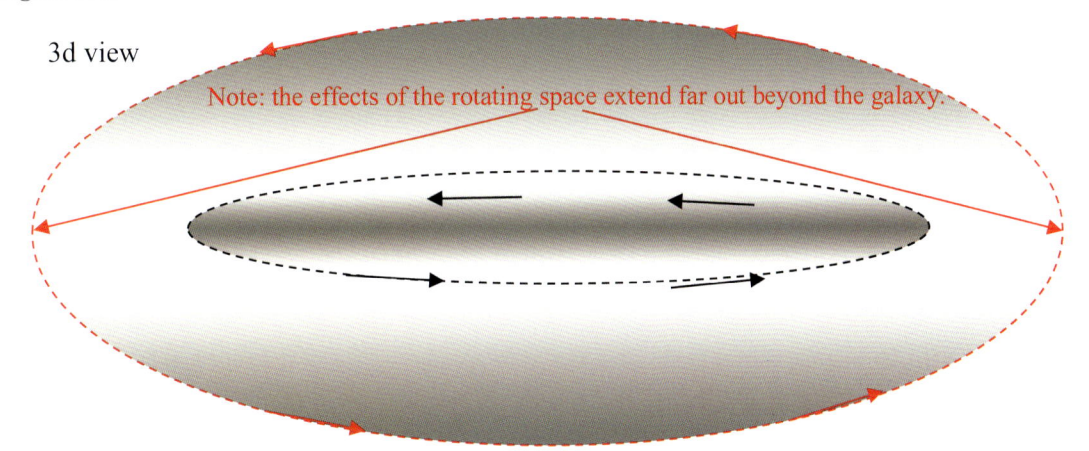

This same effect will also occur on a massive scale with galactic clusters. Smaller galaxies located close to larger more massive ones will become trapped within the rotating three dimensional space of the larger ones. Does this contribute to the "Pancake Theory" of galactic clusters? Only time and observational astronomy will tell!

Note: this same effect does not apply to individual solar systems such as ours. The sun does not generate enough of a volume of rotating space to entrap all of the planets. Only does this effect become prevalent close to the sun. Consequently, on a smaller scale its rotating space does affect the orbit of the planet Mercury that Mr. Einstein attributed to his fourth Dimension of Space-time, but this was a mistake previously exposed. Hence gravity rules the outer planets, and they all move according to Newtonian Physics.

> Dark energy is another mystery of astronomy that is again easily explained by the Vortex theory of Atomic Particles. Dark energy is…the missing anti-matter of the universe!!! It has formed anti-hydrogen, and exists in the empty space between the galaxies due to its anti-gravity effects.

Chapter 31
Dark Energy

One of the laws of the universe is the law of mutual creation. For every particle created there is an anti-particle. Using the Vortex theory of Atomic Particles we now know that this law is a function of the fact that particles and anti-particles are nothing more than the ends of tiny vortices that go into and out of 4d space. For example, the proton and the anti-proton are created thusly…

Figure 31.1

The proton and the anti-proton are connected by an invisible vortex of 3d space flowing from the proton to the anti-proton in 4d space.

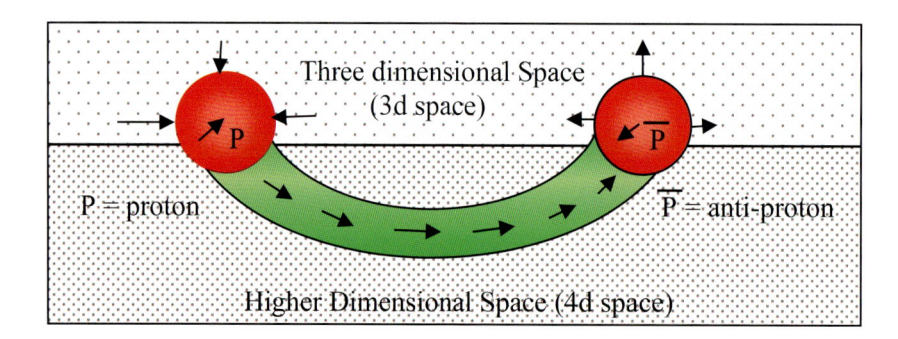

The electron and the positron are connected via a 4d vortex also…

Figure 31.2

The electron and the positron are connected by an invisible vortex of 3d space flowing from the proton to the electron in 4d space. Notice how space is flowing into the positron.

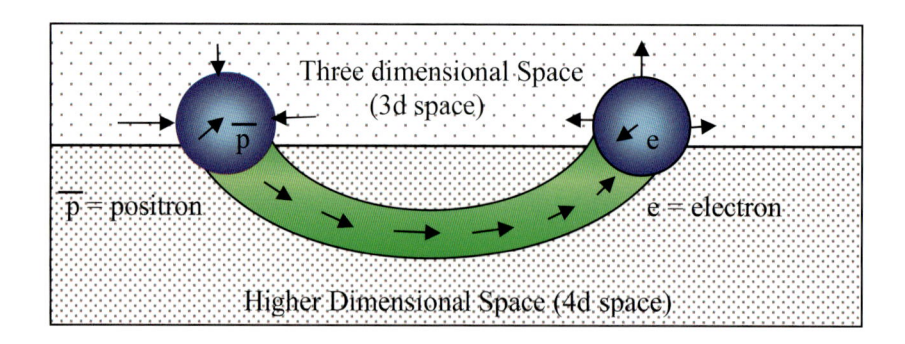

So now it is easy to see that for every proton that exists in the universe there has to be an anti-proton; and for every electron that exists there has to be a positron. We know where the protons and electrons went, they created the matter the universe is made out of. So where are the anti-protons and positrons? They could not have decayed!

The half-life of the proton is considered to be at least 1.67×10^{34} years, [that is 16,700,000,000,000,000,000,000,000,000,000,000 years!] the electrons half-life is 6.6×10^{28} years [that is 66,000,000,000,000,000,000,000,000,000 years!]. Because the universe is considered to be at least 16 billion years old; that is equal to only 16,000,000,000 years: a considerably small number compared to the one's above! So what happened to the anti-matter? [IT HAS TO EXIST!]

From observing particle collisions in linear accelerators, it is now known that only a proton can destroy an anti-proton in a matter anti-matter annihilation; and only an electron can destroy a positron in a matter anti-matter annihilation. So if all of the anti-matter was destroyed by protons and electrons, there would be no protons or electrons left in the universe because they would have all been destroyed in matter anti-matter collisions. Hence, there would be no planets, stars, or galaxies! So we are left with a problem. What happened to all of the anti-matter: enter Dark Energy.

Dark energy was discovered in 1998 by astronomers Adam Riess, Saul Perlmutter, and Brian Schmidt. The three shared the Nobel Prize for finding evidence of the accelerating expansion of the universe based on observing *Type Ia supernova* in distant galaxies. Observations that led to the hypothesis that this acceleration was caused by the presence of something between the galaxies called Dark Energy.

So what is Dark Energy? A clue to what Dark Energy is, is the anti-gravity effects it is exerting upon surrounding galaxies: causing them to accelerate away from each other.

This is a most fascinating characteristic of Dark Energy because that is exactly what anti-matter does! Although the tiny positrons are surrounded by spherical regions of space bent inward creating gravitational effects; the much more massive anti-protons are surrounded by massive regions of space bent outwards, creating anti-gravity effects!

Consequently it is now realized that in the early universe, when hydrogen was formed by the breaking of the vortices between protons and electrons forming hydrogen atoms [Figure 31.3]; the opposite would have happened too. The breaking vortices would have joined anti-protons to positrons forming anti-hydrogen. Because the majority of the bent space surrounding anti-hydrogen would have been the anti-gravity effects generated by the anti-protons, it would become anti-gravity matter.

Figure 31.3　　　　　　　　　　　　　　　**Figure 31.4**

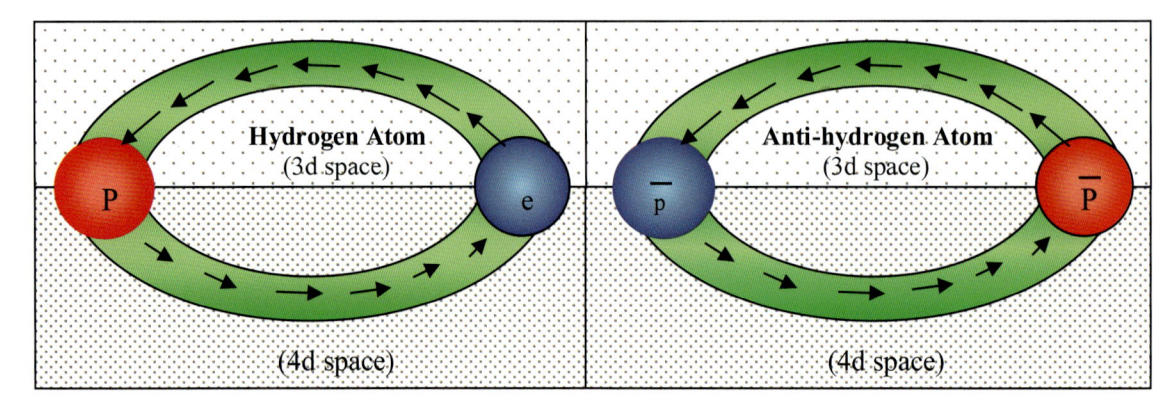

Note how the anti-proton $\overline{P}$, and the anti-electron (the positron $\overline{p}$) have joined together to form *Anti-hydrogen*…notice too how space flows into the positron and out of the anti-proton; opposite to what it does in regular hydrogen.

Figure 31.5

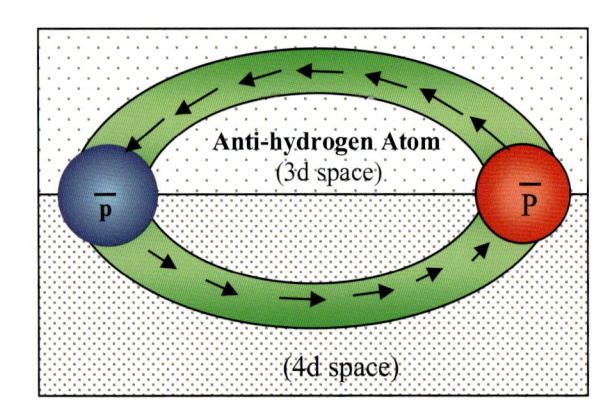

Because the anti-proton causes the surrounding space to become denser, causing the anti-hydrogen atom to create an anti-gravity effect. It cannot mix with the less dense space surrounding the proton in regular hydrogen that is surrounded by a region of less dense space. Hence, anti-matter hydrogen would be accelerated away from regular matter in the early universe. Keeping it from annihilating regular protons and electrons. It would have been regulated to the vast voids between the galaxies becoming the Dark matter of the universe.

Figure 31.6

Regular Hydrogen

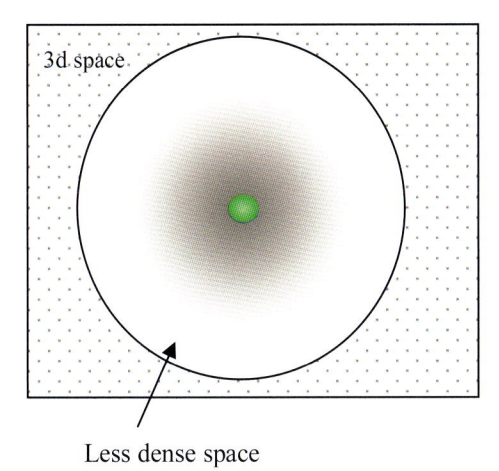

Figure 31.7

Anti-hydrogen

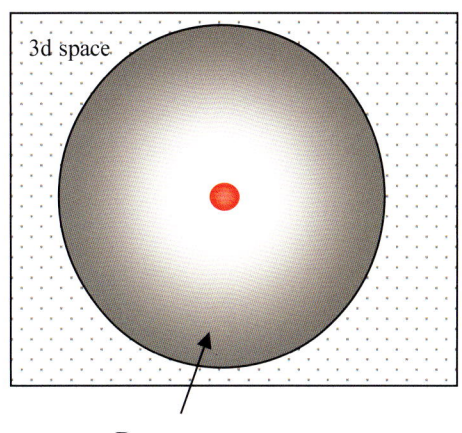

Less dense space

Denser space

    There is also an enormous amount of anti-hydrogen. The amount is equal to all of the matter in all of the 100 plus billion galaxies of the universe! Equally important, their anti-gravity effects precludes them from forming molecules as do regular atoms possessing gravitational effects. Hence, they must remain as "wandering clouds of particles"! Also, the way photons of energy pass through these clouds of anti-matter makes it seem as if the anti-matter hydrogen atoms are invisible. [They enter into positrons and exit out of anti-protons; instead of how they do when encountering regular matter: entering into protons and out of electrons], However, their presence and greater numbers, especially between the vast regions of the galaxies, is revealed by the fact that their anti-gravity force creates the acceleration of the galaxies away from each other.

Figure 31.8 The galaxies just outside the region of dark energy are accelerated away from it…

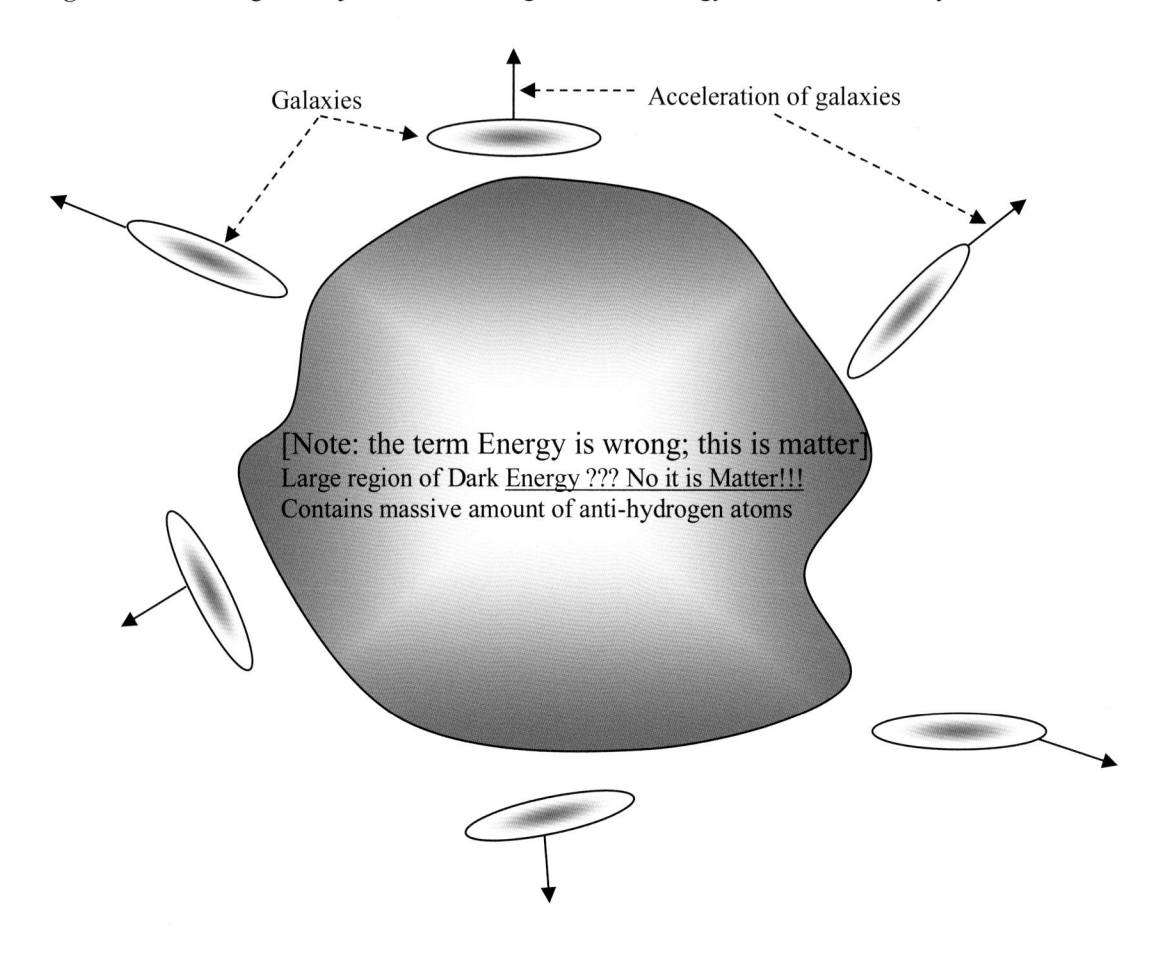

Galaxies   Acceleration of galaxies

[Note: the term Energy is wrong; this is matter]
Large region of Dark <u>Energy ??? No it is Matter!!!</u>
Contains massive amount of anti-hydrogen atoms

Two different Phenomenon created by TWO different causes!!!

Time does not exist; the phenomenon of time is created throughout the universe by the uniform flow of microscopic space into and out of the three dimensional holes of matter. This flow is uniform throughout the universe because it is created by the expansion of space itself which is uniform throughout the universe. But it is important to understand that the expansion of the space that the one gigantic particle the universe is made out of is NOT accelerating!

> The acceleration of the galaxies upon the surface of this one gigantic particle is NOT being caused by the expansion of space; but rather, by the presence of giant clouds of anti-hydrogen. The acceleration of the galaxies is an illusion that makes it appear the entire universe is accelerating. This is a false conclusion. Just as the rotation of the night sky is an illusion, so is the apparent acceleration of the universe! Both phenomenon, the expansion of space and the acceleration of the galaxies are being created by two completely different causes; one of the biggest mistakes of physics is the teaching that the transfer of force is done via certain sub-atomic particles, called boson particles. Specifically, the W, the Z, the Higgs, the Graviton, the Photon, and the Gluon! What a rogue's gallery lineup! A lineup of foolishness made by some of the best and brightest yet easily deceived men who have ever lived. What a tragedy! A tragedy that has to be rectified now…

Chapter 32
Boson Particles Do NOT Transfer the Forces of Nature!!!

One of the great tragedies of the convoluted results of reasoning with mistaken particle logic is the horrendous mistaken belief that the transfer of force has to be done by a particle. Nothing could be further from the truth.

Some Nobel prizes have been given out that should not have been! But this has to change here and now, before another generation of freshly burgeoning scientists and engineers fall victim to this great mistake! The following particles are all thieves in disguise! All masquerading with false identities…

THE ±W PARTICLE…

The –W particle below is really a breaking 4d vortex caught in a tight loop around a "0" charged "particle" such as a neutron: breaking up into a proton, electron & anti-neutrino. [The +W particle is an anti-neutron breaking up into an anti-proton, positron, and a neutrino.]

Figure 32.1 **Figure 32.2**

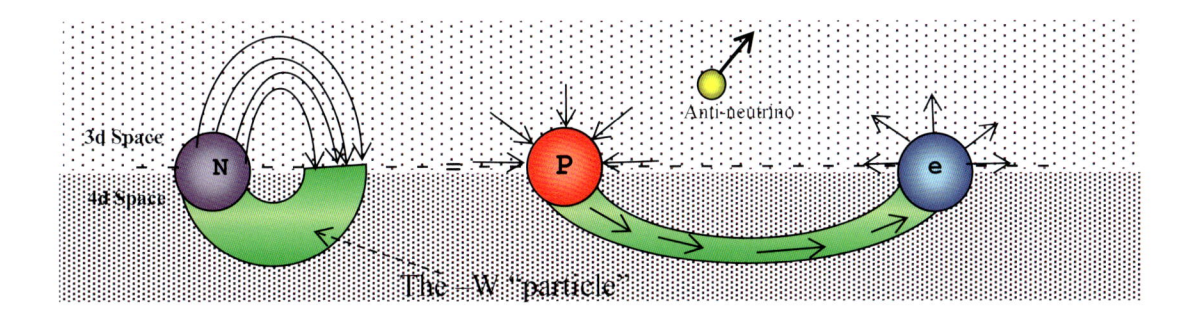

THE Z PARTICLE...

The Z particle is really a decaying 4d vortex that was caught in a tight loop that is now decaying, creating a particle anti-particle pair. In the example below, a Z particle is breaking up into an electron & positron: which are nothing more than the 3d ends of the broken loop. The collapsing vortex caught in a loop is misunderstood to be something mistaken scientists called a "Z" particle...we now see it is really a 4d vortex trapped in a loop!

Figure 32.3 The 4d loop

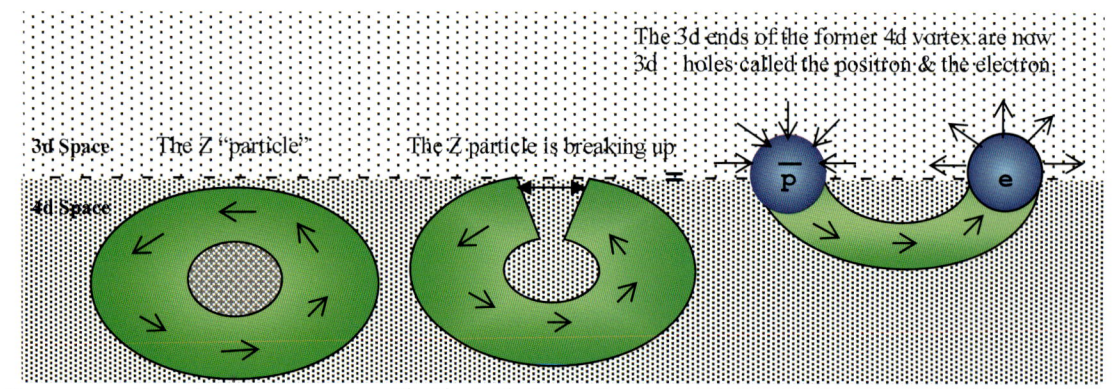

[Note: other Z particles are created in 5th, 6th, & 7th dimensional space. These higher dimensional loops decay into quark anti-quark pairs that are really their higher dimensional ends.]

THE PHOTON...

The photon is easily described as an extension of the vortex. It travels at the speed of light because the vortex it was thrown from was traveling at the speed of light. Contrary to present misunderstanding, the photon does NOT mediate or transfer the electromagnetic force. It merely elongates a flowing vortex connecting a proton to an electron; or shortens it as it is discharged. The electromagnetic force is created by a rotation of 3d space.

The photon is nothing more than a packet of condensed 3d space. The photon was discussed in Chapters 19 & 20.

THE HIGGS PARTICLE...

As mentioned before, the Higgs boson particle does not exist!

THE GRAVITON...

The graviton does not exist. It is also important to note, that when researching the graviton, it is referred to as the "hypothetical" graviton!!!

THE GLUON...

The Gluon does not exist! It does NOT hold quarks together. It is the three dimensional hole in the surface of 4d space that holds the higher dimensional holes called quarks together; not some sort of particle possessing outrageous abilities that make its force stronger as the particles it is holding together seek to move further apart! This is just another mistaken, fantasy, & false assumption made using the erroneous "reasoning process called Particle Logic"!

Figure 32.4   **Figure 32.5**

It is the 3d hole that holds the 4d holes [quarks] together, not some sort of exotic particle called the gluon that "glues" them together.

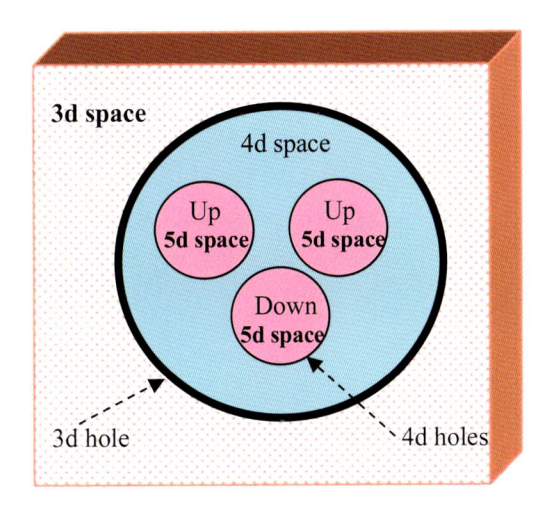

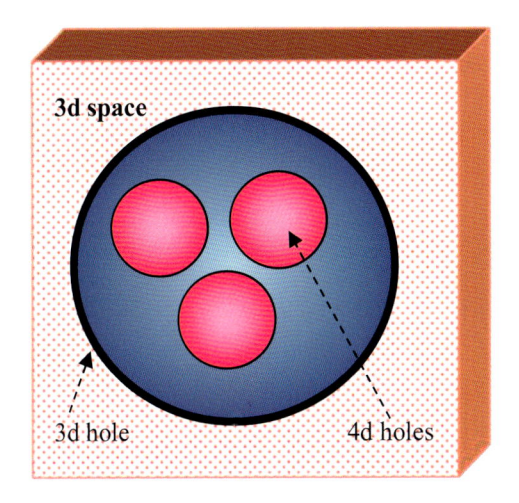

Dispelling the Myth of Quantum Field Theories

"Field Theories" were invented to try to explain the phenomenon that space was made of nothing, could not explain. Note: how can you have a field that supposedly permeates the entire universe yet violates the very principle of a field: no point of origin; and does not dissipate with the square of the distance. It is a fantasy, created in the minds of desperate scientists!

> The mystery of how particles can miraculously interact at great distances from each other can now be explained. Using the principles of the vortex theory it is easy to see how quantum entanglement creates quantum vortices that are responsible for the interaction taking place between particles separated by great distances.

Chapter 33
The Explanation of Quantum Entanglement

4D & 3D VORTICES ARE WHAT CAUSES QUANTUM ENTANGLEMENT

The great mystery of quantum entanglement is now solved. The invisible vortices connecting "particles" together solves the problem.

Einstein did not like quantum mechanics because it suggests everything in the universe is ruled by the laws of probability. He attacked quantum mechanics by pointing out it would not explain quantum entanglement. He stated that this failure suggested that quantum mechanics was an incomplete theory and that there had to be "Hidden Variables" responsible for the seemingly random measurement results. These assumptions about the existence of hidden variables have come to be known as "local realism". But not anymore!

Until now, these assumptions about the existence of hidden variables were only assumptions. But with the discovery of The Vortex Theory they are now revealed.

It is too bad that the last Century's giants of Physics, Einstein and Niels Bohr who argued about the origins of Quantum Entanglement did not know this. Otherwise there would have been no conflict between them. If they would have seen the below illustration, the conflict would have been solved before it began. Nor would there have been any reference to Einstein's "Spooky action at a distance."

Figure 33.1

In the drawing below, the proton and the anti-proton are connected by an invisible vortex of 3d space flowing from the proton to the anti-proton in 4d space. This vortex Quantum Entangles them...

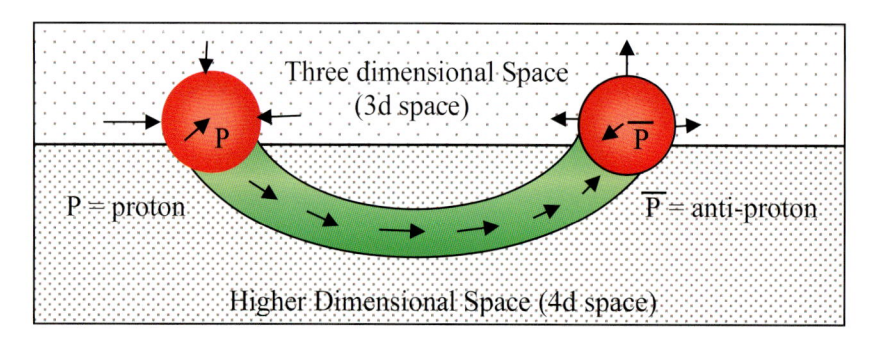

The dynamics of this vortex are most impressive. It is akin to a pipe. Try this experiment, take a short length of pipe and twist it counterclockwise (CCW). Notice how the whole pipe rotates. Also note how a CCW rotation at one end causes a CW rotation in the other end. This happens because it is easy to see the two holes are the ends of the pipe.

This same phenomenon happens no matter how long the pipe is. If it is a mile long, the opposite end will immediately begin to rotate when the other end is rotated. The same exact situation happens with the proton anti-proton pair. Rotate one end in an electromagnetic field and the opposite end will rotate too. This situation occurs because the proton and anti-proton are merely the ends of the same vortex. Like the pipe, rotate one end and the other rotates too, no matter how far they are separated.

This phenomenon occurs because the rotating vortex rotates as a single entity. One object!

THE 3D VORTICES ARE WHAT CAUSES QUANTUM ENTANGLEMENT IN PHOTONS

If two photons possessing complementary polarizations (opposite Spins) touch and become "entangled", this extreme closeness forces their rotations extending into three dimensional space to join together forming a vortex. When the particles are separated, the vortex between them remains. A change in the orientation of a photon at one end of the vortex travels in a wave down the length of the vortex creating a change in the orientation of the photon at the other end.

At the time of this writing, it is not known if the transference happens instantaneously or at the speed of light. Only experimentation will tell.

Figure 33.2A Top view   **Figure 33.2B** Side view

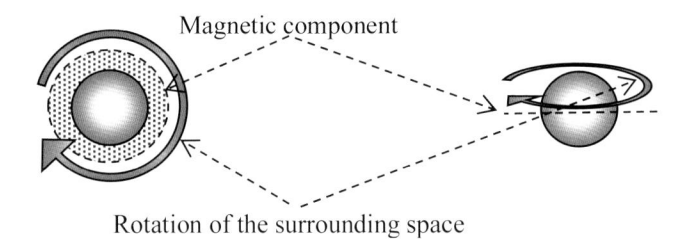

Magnetic component of a particle is created when its rotation rotates the surrounding space.

PHOTON QUANTUM ENTANGLEMENT

Figure 33.3

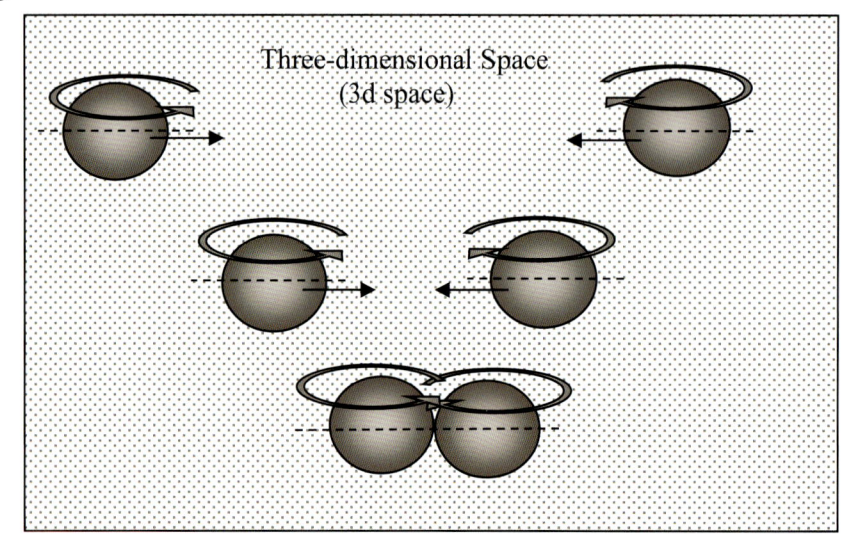

When the two photons touch, their opposite spins cause the 3d space between them to begin to rotate creating a vortex of spinning space. As the two photons move away from each other, the vortex increases in length. How far will it go? Only experimentation will tell.

Figure 33.4

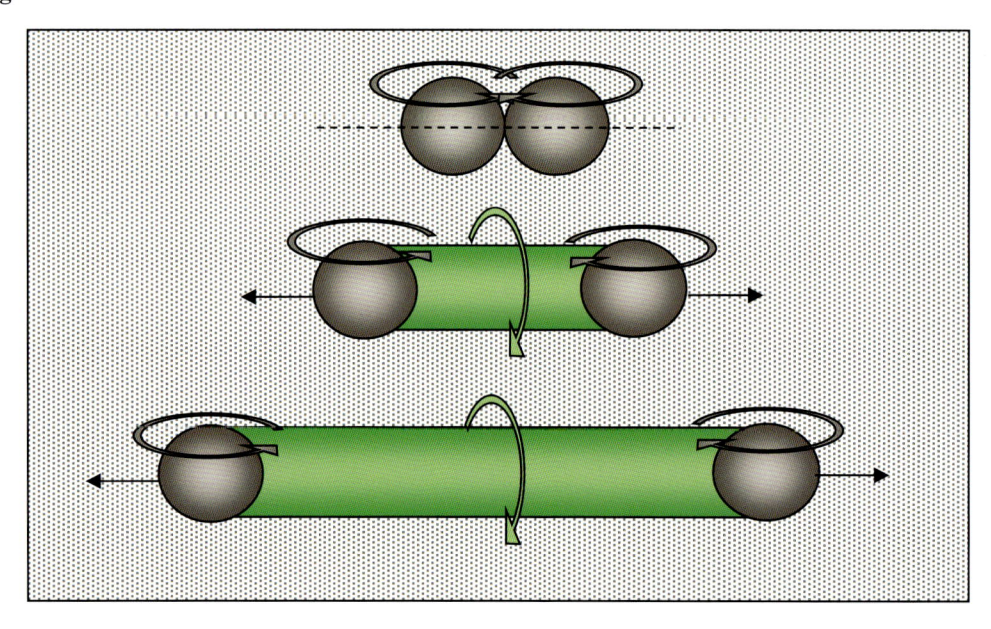

This joining together of the two vortices creates a vortex that is different from the vortex joining the proton and the electron. The vortex joining the proton and the electron can be envisioned [from our point of view], as a whirling hollow tubular shaped structure [similar to one formed above the drain in a sink]; whereas the vortex joining the two photons is merely a vortex of whirling space [similar to a tornado in the atmosphere]. **They now expand and contract together.**

When the two entangled photons are separated by a beam-splitter, and they move off in opposite directions, the vortex now called the "Quantum Vortex" elongates:

Unlike vortices we are used to seeing in the oceans and the atmosphere, the vortices in this three dimensional space are not created out of whirling particles. Vortices created out of whirling particles are limited in length due to the momentum of the particles they are created out of.

These vortices in fourth dimensional space are created out of space itself. Consequently, as the vortex elongates, there is no loss of angular momentum due to such factors as friction limiting its length. Hence, the vortex is unlimited in length unless it crosses another vortex or is affected by perturbations at the ends.

QUANTUM TELEPORTATION:

There are two explanations for the transmission of the effects of quantum teleportation. Only experimentation will determine which one is correct.

Explanation #1

The quantum teleportation effect that takes place between two previously complementary and entangled photons is first illustrated in the following way:

Imagine two men holding either end of a rope stretched between them. Now imagine one of the men suddenly and swiftly raising and lowering his hand. This action sends a wave down the rope towards the hand of the man at the other end of the rope; and as the wave reaches his hand, it too is suddenly raised and lowered.

A similar situation occurs between the two previously complementary entangled photons.

Figure 33.5

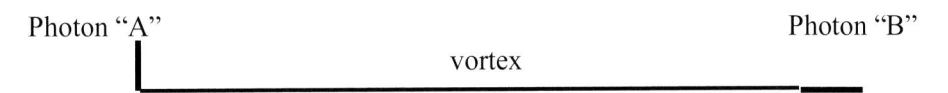

```
Photon "A"                                                    Photon "B"
    |_____vortex_____|
```

In Figure 33.5, two photons labeled "A" and "B" are connected to each other by the Quantum Vortex. [The two photons are traveling along the positive "Z" axis perpendicular to the page and moving in a direction directly towards the reader.]

The photons remain in this orientation unless outside forces create a change.

A change is created when the vertical vibration of orientation of photon "A" is suddenly changed by a quantum entanglement with a third photon [call it "C"].

If "C" is at a 45 degree angle to "A", their electromagnetic characteristics are also at 45 degree angles to each other. This mutual orientation causes them to rotate, causing the vertical orientation of the photons to rotate; causing the vertical orientation of photon "A" to suddenly shift 45 degrees.

This sudden 45 degree change in the orientation of photon "A" creates a wave that travels down the length of the quantum vortex to the "B" photon changing its orientation 45 degrees.

Unlike the example of the two men and the rope, the wave would not be reflected back again towards the "A" photon. This reverse effect does not happen because unlike the rope, the photons are not connected to another structure [such as the hand of the man] that retransmits the wave back down the length of the vortex.

The entire process is seen in Figures 33.6 to 33.9:

Figure 33.6

Photon "A" is suddenly shifted 45 degrees.

Photon "A"   Photon "B"

Figure 33.7

As photon "A" suddenly twists in space, a wave is created in the vortex.

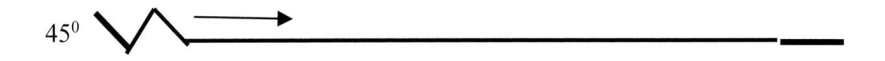

45⁰

Figure 33.8

The wave travels down the vortex at the speed of light.[5]

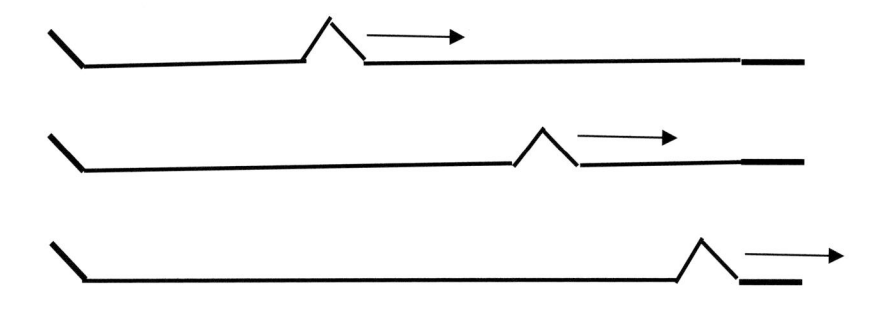

Figure 33.9

When the wave arrives at photon "B", its orientation is twisted 45 degrees.

Photon "A"   Photon "B"

45⁰

---

[5] Note: because the magnetic field rotates at the speed of light, and since the movement of the magnetic field is responsible for the shift in the orientation of the "A" photon, the shift in the orientation would also occur at the speed of light. Making the wave travel down the vortex at the speed of the shift.

Explanation #2

The second explanation of quantum teleportation is identical to the first with the following exception: the wave does not travel down the vortex. Instead, the entire vortex acting as a single structure shifts its orientation causing the photon at the other end to shift its orientation. It would happen thus:

The vortex can be envisioned as a long tube of spinning space forming a single structure:

Figure 33.10A

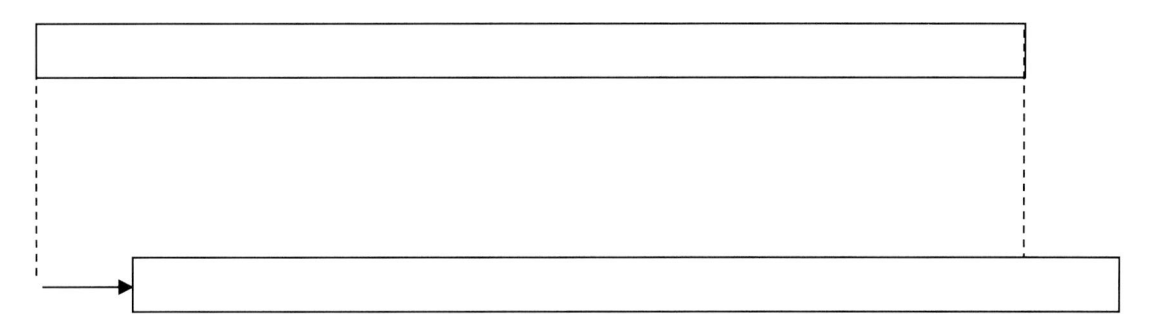

If it is indeed a "single structure" such as a long plastic tube, it would then move as a single unit. Consequently, pushing on one end changes the position of not only that end but the position of the entire tube. Creating for all intents and purposes, an instantaneous change in position of the other end.

Such a structure would also create another seemingly instantaneous effect: the distortion in one end would cause a corresponding change in the other end. For example, if a force was applied at point "b" in such a structure, it would deform inward creating a long thin parallelogram

Figure 33.10B

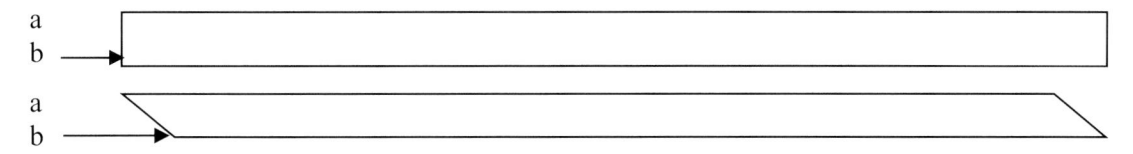

If this is the way in which the vortex is constructed, anything that affects the shape of one end [photon A] will appear to instantaneously affect the other end [photon B]. The same holds true for the explanation of the effects created by any two particles such as electron-electron entanglement.

"ELECTRON – ELECTRON" ENTANGLEMENT…

The quantum vortices of the photon, photon pair were created by their rotation of three dimensional space. Because this unique characteristic is also possessed by other "particles" such as the proton, electron, and the neutron; they will also be able to create quantum vortices. These vortices will then enable them to create the quantum teleportation effect.

Because a most ingenuous and successful experiment involving electrons was conducted in 2006 by Robert Desbrandes and Daniel Van Gent, quantum teleportation using electron, electron effects as the example are examined.

During a brief examination of the vortex theory's basic principles, it does not appear as if two electrons can be entangled. Since they are already connected to other particles by fourth dimensional vortices [represented by Figure 33.1], the existence of yet another vortex connecting two electrons together does not seem possible. However, a closer inspection of the relationship between three dimensional space and higher dimensional space allows for the possible existence of another similar though differently constructed vortex.

In the figure below, two electrons in the outer shells of two atoms pressed up against each other approach each other. Normally they cannot touch each other because their electrostatic charges are too powerful. However, if they exist upon the shell of an atom, their charges are flowing back to the protons in the nucleus, and it is easier for them to make contact.

Figure 33.11

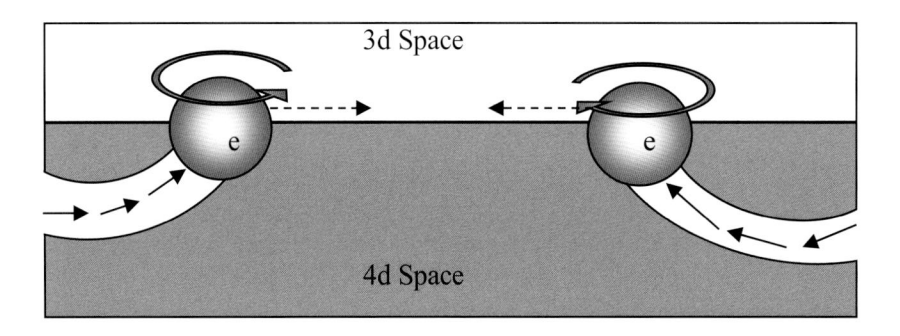

In the figure below, the two electrons in sub-shells of two close atoms approach each other:

Figure 33.12

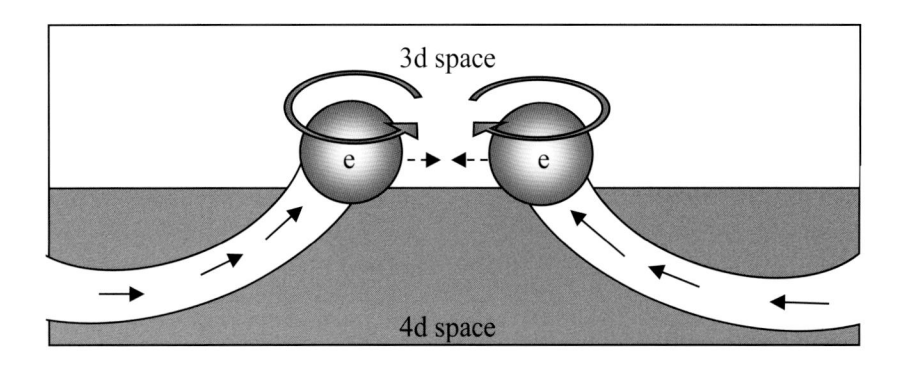

Consequently, since the spin states of the two electrons are opposite to each other [spin up and spin down] and if they are somehow pressed close enough to each other, like the photons in Figure 33.14, their rotations can begin to create a second vortex of spinning space between them in 3d space.

Figure 33.13

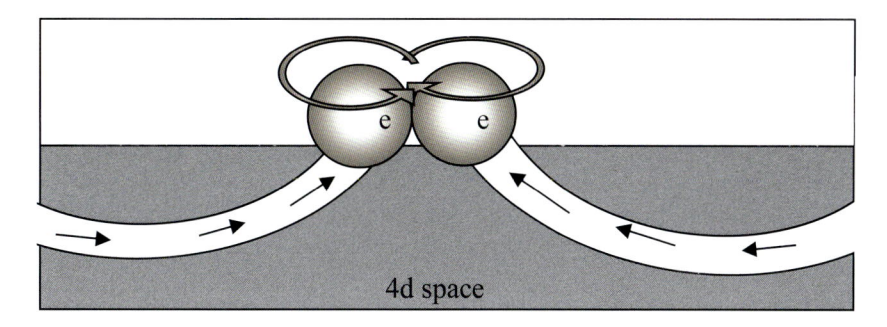

And when the two 3d regions of space surrounding each electron vortices join together, they create a vortex of whirling space between them that remains and elongates as the two electrons disengage.

Figure 33.14

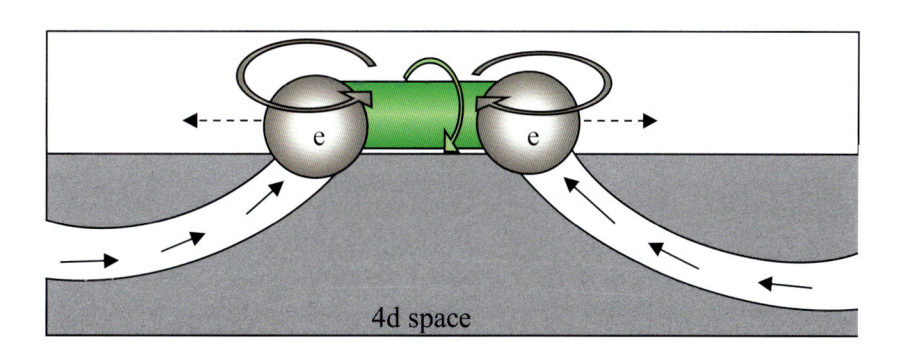

Again, it must be remembered that the vortex between the two electrons in 3d space can be envisioned as a whirlwind of space [similar to a tornado], whereas the vortices connecting the electron to a proton or positron in 4d space can be envisioned as a whirling elongated tube similar to a rotating pipe.

Consequently, when this quantum vortex is created, a link now exists between the two electrons in 3d space that is similar to the one that existed between the two photons. However, this quantum vortex is larger and stronger than the one existing between the two photons because it was created by the much stronger electromagnetic fields surrounding the two electrons.

QUANTUM TELEPORTATION EFFECTS BETWEEN ELECTRONS:

The quantum teleportation effects that occur between two entangled electrons is similar to those created by the two photons. When one of the electrons is suddenly agitated say by a burst of photons, its sudden motion or vibration will send waves down the 3d vortex causing the other electron to be similarly agitated. Or, if one electron is caught in a magnetic field and flipped, the effect will travel down the vortex and the other one will flip too.

ENTANGLEMENT SWAPPING

The mystery of entanglement swapping is easily explained when it is realized that entangled electrons and protons are connected by quantum vortices.

As seen by classical science, two pairs of entangled particles [A & B] and [C & D] appear to be separate entities, existing alone and apart from the other particles of the universe:

Figure 33.15

However, from the viewpoint of the vortex theory, they are not separate but joined by fourth dimensional quantum vortices:

Figure 33.16

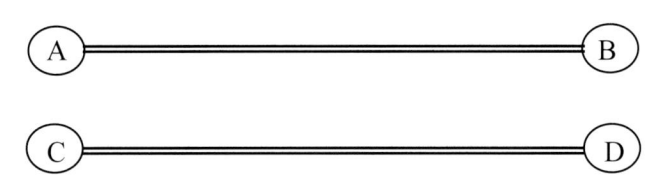

If particles B and D are then subsequently entangled the vortices will break, and connect to each other, entangling A and C together:

Figure 33.17

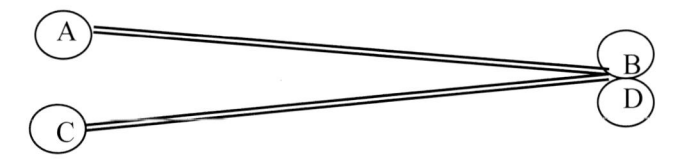

Figure 33.18

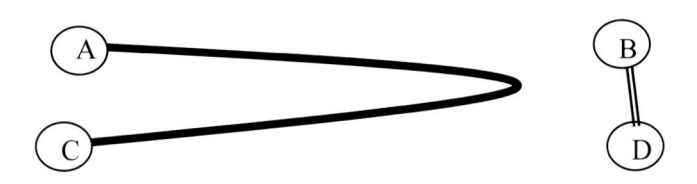

Particles A and C are now connected to each other leaving B and D separate and isolated but connected to each other. It is easy to see how B can no longer interact with A, and D can no longer interact with C; but A and C are now able to interact with each other.

CONCLUSION:

The mystery of how particles can miraculously interact at great distances from each other can be explained. Using the principles of the vortex theory it is easy to see how quantum entanglement creates quantum vortices that are responsible for the interaction taking place between particles separated by great distances.

Albert Einstein was absolutely correct in his famous paper published with Podolsky and Rosen. His view that quantum mechanics was an incomplete science and that there had to be hidden variables responsible for seemingly random events was inspired intuition. Unfortunately, his vision of the universe sees space to be made of nothing, and this is where the problem lies. His *postulate* that space is made of nothing does not allow one to deduce the possible presence of vortices responsible for the phenomenon of quantum teleportation.

> It is one thing to draw the sub-shells of atoms, it is another to understand the real cause of how they are created. The Vortex Theory of Atomic Particles now makes this wonderful explanation possible.

Chapter 34
How the Four Quantum Numbers of Atoms Are Created

Twentieth Century Physics states that using the solution to Schrödinger's equation, a set of complicated mathematical equations are obtained called wave functions, which describe the probability of finding electrons at different locations within an atom. These wave functions describe shapes called atomic orbitals. These atomic orbitals in turn describe regions of space in which there is a high probability of finding an electron. These regions are designated by four quantum numbers of which no two electrons can have the same ones. Consequently, we are taught to believe in chemistry classes that it is the quantum numbers that dictate what locations the electrons will be located in as they surround the atom. But what these quantum numbers are really describing are the final effects and not the initial cause.

When I originally began to write this chapter, the complexity of the nucleus of the atom began to make it longer and longer. Consequently, I realized the uncomfortable truth that this information has to be put into a pamphlet all its own. So I will paraphrase here the explanations of the four quantum numbers and then later, put out the pamphlet for chemistry students explaining the simple explanations for the exceedingly complex orbitals of the heavier nuclei and many more phenomenon such as the decrease in diameter of atoms in the periodic table as we go from left to right; and the increase in size as they go from top to bottom.

THE FOUR QUANTUM NUMBERS

Principal Quantum Number (n): $n = 1, 2, 3$, etc. Defines the distance from the electron to the nucleus and is easily explained by the elongation of the vortex. [According to chemistry, the greater the length of the distance to the nucleus, the greater the energy.]

Figure 34.1

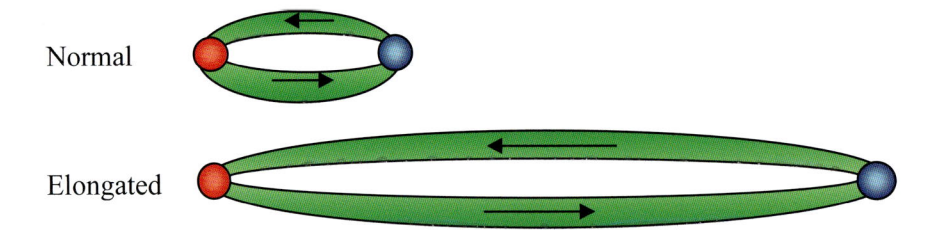

Normal

Elongated

It is easy to see that the longer vortex possesses a greater volume of 3d space in its two vortices. Since this 3d volume of space represents energy, it is easy to see this configuration possesses more energy.

The Angular Momentum Quantum Number (l): l = 0, n-1. Specifies the shape of an orbital with a particular principal quantum number.

The key to some of these shapes in larger atoms is the limited degree of motion of the vortices flowing from an electron to the proton's location in the nucleus of the atom. The lesser the degree to which they can move from side to side limits the ability of the electron to move; entrapping it in the volume of space that has come to be called the orbital.

These shapes are created by four rules:

#1 the position and location of the alpha particle in the nucleus; or the single proton neutron pair the electron's 3d vortex is connected to in the nucleus, and its length;

#2 the position where the electron's 3d vortex penetrates into the surface of the nucleus & either limits or engages the electron's ability to rotate;

#3 condensed, enlarged 3d & 4d vortices caused by the large volume of less dense space in larger atoms, that spread the alpha particles apart, and make the nucleus unstable;

#4 the electron's denser region of 3d space that surrounds it, pushing against the other denser regions of space surrounding other nearby electrons;

Figure 34.2   Examples of orbital shapes…

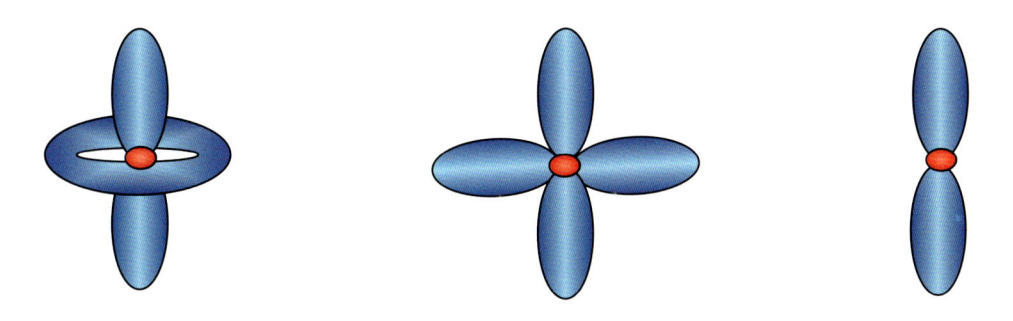

▶Important observation:◀

It is certainly feasible for the electron to exist in the very end portion of the orbital seen in Figure 34.2A. However for it to stray into the lower section seems unfeasible. For this to happen both the 3d & 4d vortices would be forced to greatly shorten. This would cause the proton and the electron to suddenly and dramatically increase in size; causing the alpha particle in the nucleus to become unstable.

Figure 34.2A   **Figure 34.2B**

Maybe   **Unlikely?**

Electron's location in a sub-shell?

Nucleus

Magnetic Quantum Number (ml): ml = -l, 0, +l.
Defines the orientation in space of an orbital of energy (n) and shape (l). These shapes are created by four rules…

Rules #1, #2, #4 and the following…

#3 condensed, enlarged 3d & 4d vortices caused by the large volume of less dense space in larger atoms, that spread the alpha particles apart, and make the nucleus unstable;

#5 the amount of less dense 3d space the atom possesses, causing shrinkage of the lengths of the vortices;

#6 how the alpha particles and the other protons and neutrons in the nucleus of the atom are stacked on top of each other, beginning with hydrogen and ending with Oganesson;

Figure 34.3

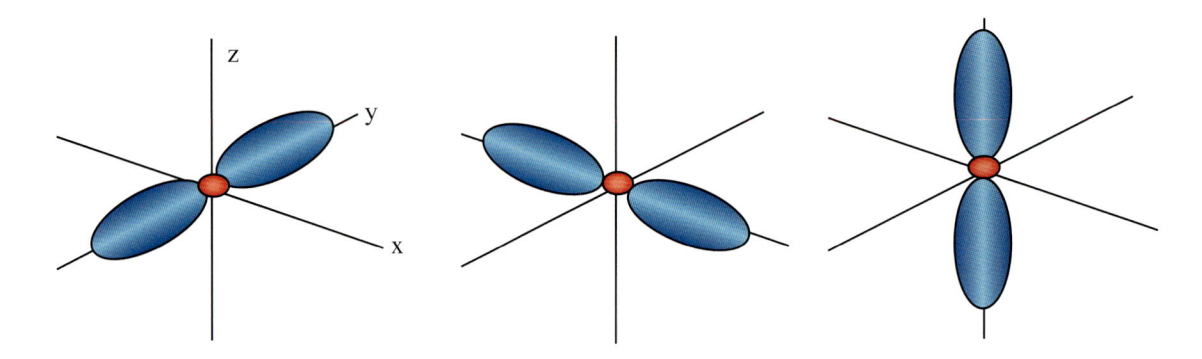

Spin Quantum Number (ms): ms = +½ or -½.
Defines the orientation of the spin axis of an electron. An electron like a tiny magnet can only spin either "up" or "down". This effect is being created by the alpha particle in the nucleus as explained by the Pauli Exclusion Principle…From Book 3, Ch 25 The explanation of the Pauli Exclusion Principle…

Surprisingly, using the principles of the Vortex Theory, the explanation for Pauli Exclusion Principle can now be illustrated; [although the spin states of each type of particle is opposite, note particularly the spin states of the electrons]: These shapes are created by two rules...

#7 the real explanation for the Pauli Exclusion Principle: [caused by the opposite spins of the 2 protons the 2 electrons are connected to in the alpha particle];

#8 and the opposite spins of the two electrons via the Pauli Exclusion Principle in a sub-shell that attract them and cause them to combine into Cooper pairs; & make their vortices the same length.

Figure 34.4

ALPHA PARTICLE

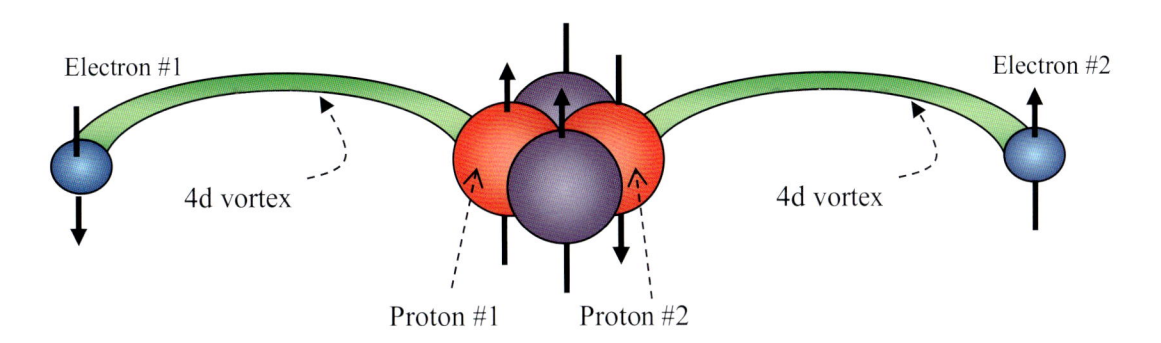

Note how the spin states of the protons in the alpha particle are opposite to each other. Although the spin states of the proton and the electron it is attached to are also opposites. Because each is at the other end of the rotating vortex, the spin states of the two electrons are now opposite to each other; consequently, ***the spin states of the two electrons are opposite to each other because the spin states of the protons they are attached to are opposite to each other.*** [Note: this is not conjecture. We know that the spin states of the two protons and the two neutrons are opposite because the spin states of alpha particles are zero. Indicating they are all canceling each other out. This is the simple yet elegant explanation for the Pauli Exclusion Principle.

Figure 34.5

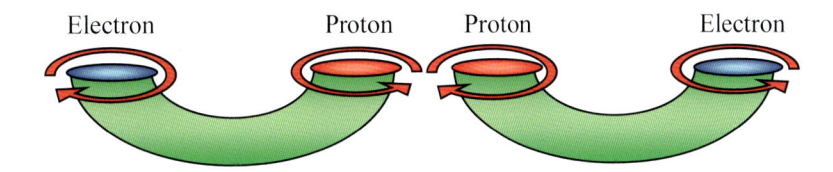

However, if the atom has an odd number of protons, one of these will be a proton neutron pair. This will create either one of two configurations:

Figure 34.6 Electron spins down…caused by proton's spin up…   **Figure 34.7**

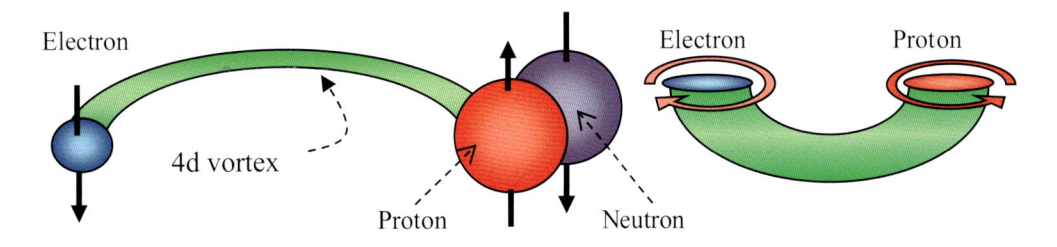

Figure 34.8   **Figure 34.9** Electron spins up…caused by proton's spin down…

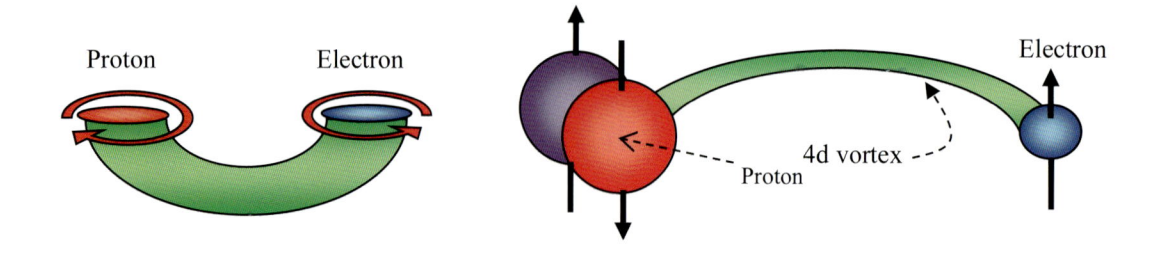

> Radioactive Decay is explained when it is realized that if an electron - electron pair is forced too close to the nucleus of the atom, the vortices are compressed causing the alpha particle they are connected to expand in size and be expelled out of the nucleus.

Chapter 35
Radioactive Decay

RADIOACTIVE DECAY

In the ▶Important observation:◀ in the last chapter, it can be seen that it is unlikely that the electron will migrate to the bottom half of the sub-shell. To do so means the vortices would have to be compressed, increasing the size of their 2 ends: the protons and electrons.

However, if the sub-shell is filled by two electrons, and they are indeed forced to the bottom of the sub-shell by say the surrounding electrons being in positions where their combined denser space forces the two electrons downwards towards the nucleus, the inward crush of the two compressed vortices on the alpha particle in the nucleus would make it want to suddenly swell to an enormous size.

But if it does, the close proximity of the surrounding alpha particles crush on it would not allow this to happen. So the only option is for the alpha particle to tunnel into 4d space, move out of the nucleus, and then return to 3d space once it is outside of the atom.

Figure 35.1

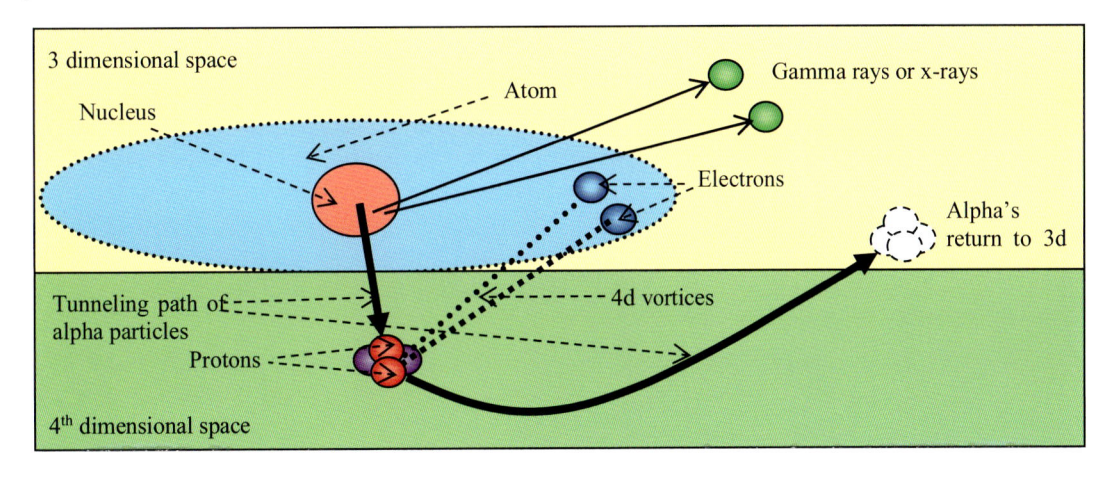

When the alpha particle returns to the three dimensional surface, its two protons are still connected to their two electrons via their fourth dimensional vortices. However, they are no longer connected to the electrons via their three dimensional vortices. These volumes of three dimensional space that once connected the electrons to the protons, continued to flow towards the position in the nucleus where the two protons were; then out of the nucleus and into three dimensional space; becoming two gamma rays.

However, if the electrons were in shells closer to the nucleus, their vortices would be shorter, and the volumes of the photons would be less. Hence, instead of gamma rays being emitted from the nucleus, x-rays would be emitted. Either way, the freed unattached electrons that are left, now give the atom a charge of – 2, [that is if they are not also thrown free]; while the 3d space flowing back into the protons from the electrons gives the alpha particle its + 2 charge.

Again, it is the sudden enlargement and instability of an alpha particle that is the reason why some more massive atoms tend to decay by ejecting alpha particles.

It is theorized that harmonic motions of the electrons in the surrounding sub-shells create resonances that cause different half-life's for different elements.

> Beta Minus and Plus decay is absolutely fascinating. The Vortex Theory not only describes how it happens but introduces two new revolutionary particles to the sub-atomic world: the Tunneling Positive Pion; and the Tunneling Negative Pion.

Chapter 36
Beta Minus & Beta Plus Decay

BETA MINUS DECAY…

There are two types of beta decay: *beta minus* and *beta plus*. For beta minus (β⁻) decay, a neutron is converted to a proton, an electron and an electron antineutrino. The electron [beta particle] and the Anti-neutrino are ejected out of the atom.

Figure 36.1 Nucleus

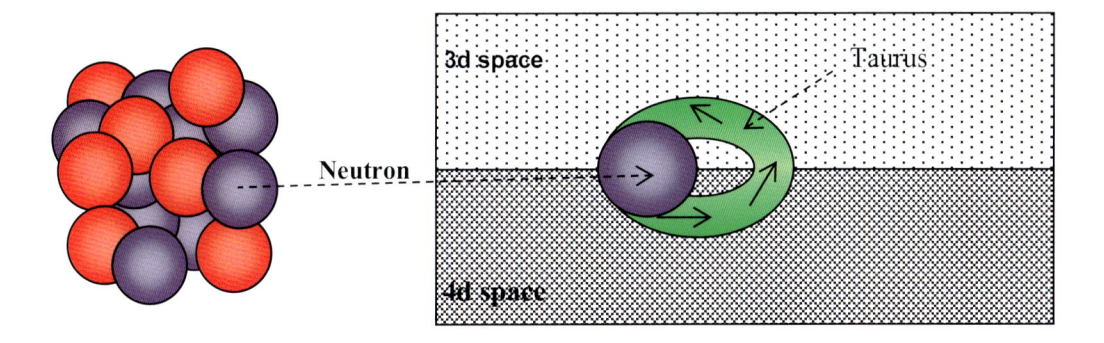

Figure 36.2 Vortex breaks

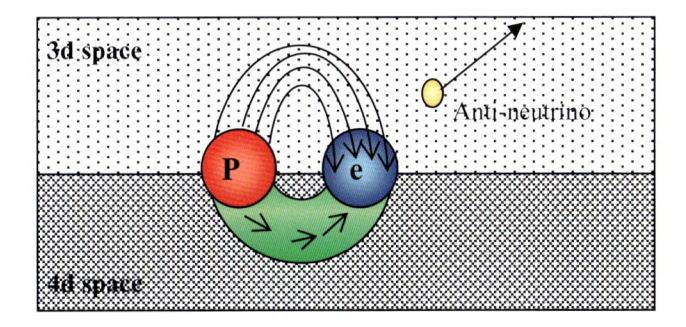

Note how the vortex breaks into its constituent parts: revealing one end to be the proton, and the other to be the electron. The anti-neutrino is the extra volume of space that was trapped within the neutron's larger volume.

Figure 36.3

As the proton and the electron move away from each other, the 3d space again begins to flow into the proton and out of the electron.

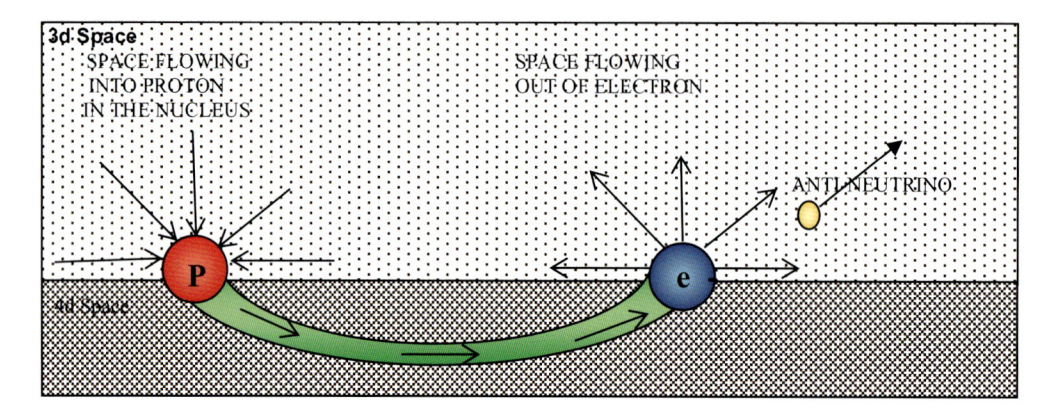

HOW DO THE QUARKS CHANGE FLAVOR?

The quarks change flavor in a most unusual and revolutionary way, they create a new particle in nature that the Vortex Theory of Atomic Particles designates as *The Tunneling Negative Pion!* Here is how it happens…

> Below are SCHEMATIC DRAWINGS developed & explained in Books 3.

DISCUSSION

The explanation of the creation of the new undiscovered particle in nature, the **Tunneling Negative Pion** begins with the discovery from Books I & II of the two sides of space, [just as a piece of paper has two sides, so does space]. Here it is explained that Up [**U**] quarks are formed on one side of space, and Down [**d**] quarks are formed on the other side of space. It can also be seen in the figure below that the Up is connected to the Anti-Up [$\overline{U}$] in an Anti-proton somewhere in the universe and the two Downs [**d**] are connected to two Anti-Downs [$\overline{d}$] in Anti-particles somewhere in the universe.

Figure 36.4   NEUTRON

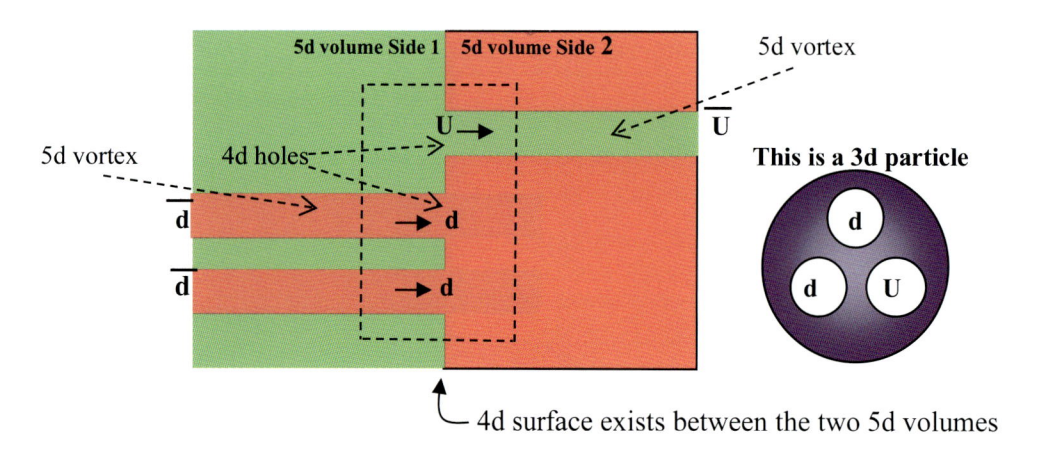

Figure 36.5

When the Neutron breaks up as seen in Figure 36.2 above, in the figure below, several things occur simultaneously: although impossible to draw, the unfolding 4d surface pulls against the side1 volume and an Up, Anti-up pair is created.

NEUTRON IN TRANSITION

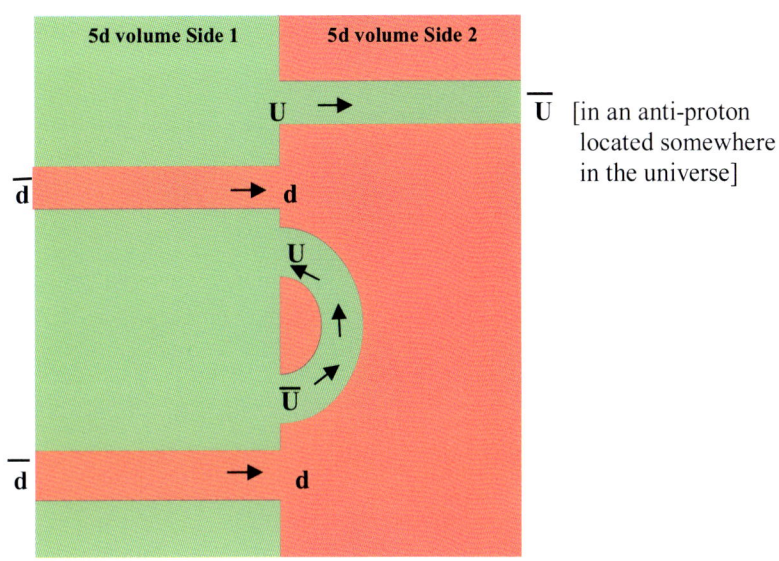

$\overline{U}$ [in an anti-proton located somewhere in the universe]

When it is all over, the Up quark in the new Up Anti-up pair now connects to the existing Up quark and Down quark creating the Proton; while the Anti-up quark connects to the other Down quark and they create a **Tunneling Negative Pion** that moves away from the proton in fourth dimensional space, invisible to our eyesight.

In the figure below, two particles are created: the **Proton** and the **Tunneling Negative Pion**.

Figure 36.6   Creation of Proton and Tunneling Negative Pion

$\overline{d}$ = Anti-down quark
$\overline{U}$ = Anti-up quark
d = Down quark
U = Up quark

This is a 3d particle!

This is a 4d particle!

135

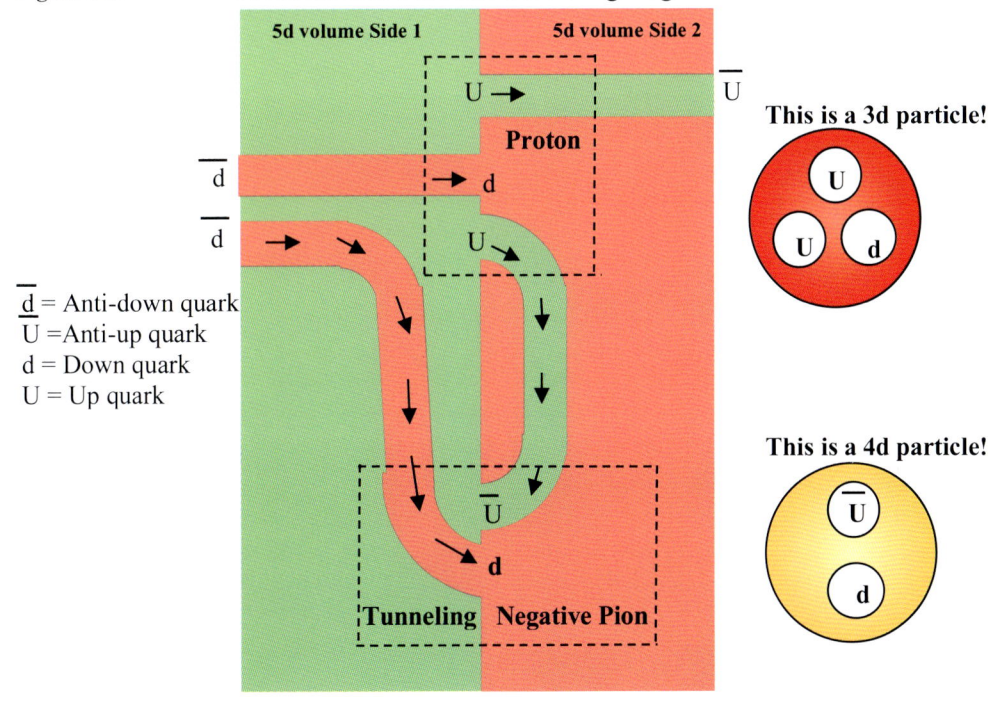

The **Tunneling Negative Pion** moves away through 4d space, where it is invisible to 3d instruments.

BETA PLUS DECAY…

For beta plus (β⁺) decay, a proton is converted to a neutron in the nucleus and emits a positron.

Figure 36.7 Nucleus

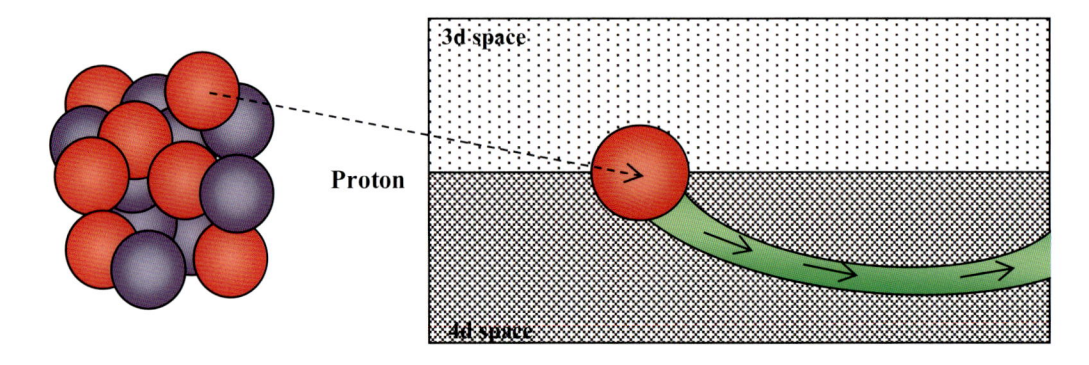

In the figure below, the proton's 4d vortex breaks…

Figure 36.8

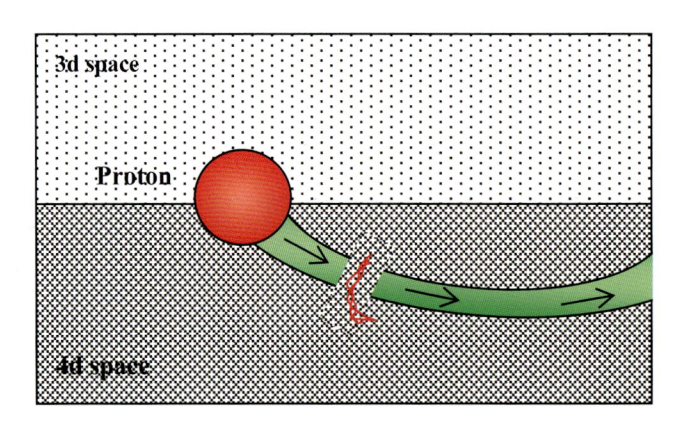

When the vortex breaks, one end curls back towards the proton, while the other end heads towards the 3d surface. Notice too, when the vortex breaks, observe how the one end about to encircle the proton, and the other end, if rejoined, would still be flowing away from the proton as seen in the figure above. However, because they both are broken, and reach the surface, one appears to flow towards the surface [creating a negative charge] while the other flows away from the surface [creating a positive charge]. This is yet another example of how the vortex creates the phenomenon known as Conservation of Charge.

Figure 36.9 Vortex breaks

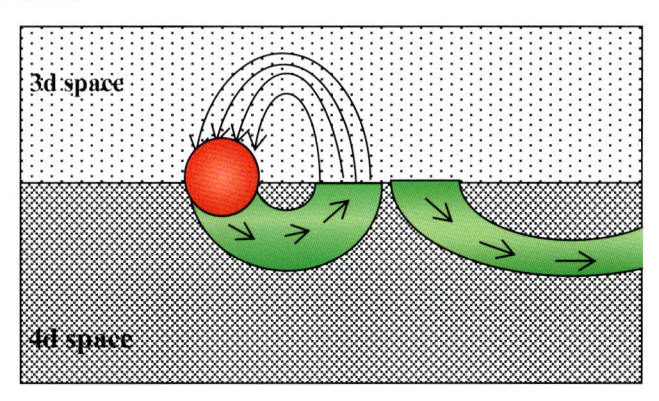

The end that wraps around the proton becomes the neutron, and the other end becomes a positron. NOTE: this is only a SCHEMATIC DRAWING. In reality, the wrapping of the vortex around the proton creates a 4d torus that is impossible to draw

But again, what happens to the quarks?

HOW DO THE QUARKS CHANGE FLAVER?

Again, the quarks change flavor in a most unusual and revolutionary way, they create another brand new particle in nature the Vortex Theory of Atomic Particles designates as *The Tunneling Positive Pion!*

Here is how it happens…

DISCUSSION

The explanation of the creation of another new undiscovered particle in nature, the **Tunneling Anti Positive Pion** begins again with the discovery from Books 1 & 2 of the two sides of space. Here it is explained that the two Up [U] quarks are formed on one side of space, and one Down [d] quark is formed on the other side of space. It can also be seen in the figure below that the Up is connected to the Anti-Up [$\overline{U}$] quark in an Anti-proton somewhere in the universe; and the Down [d] is connected to an Anti-Down [$\overline{d}$] quark also in an Anti-particle somewhere in the universe.

Figure 36.10

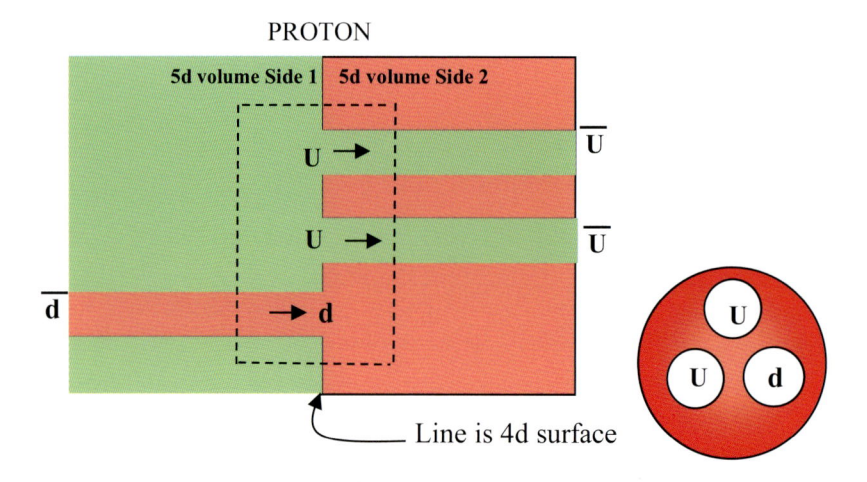

In the figure below, when the Proton breaks up the enveloping vortex *pulls backward on side 2*, creating a Down, Anti-down pair of quarks.

Figure 36.11

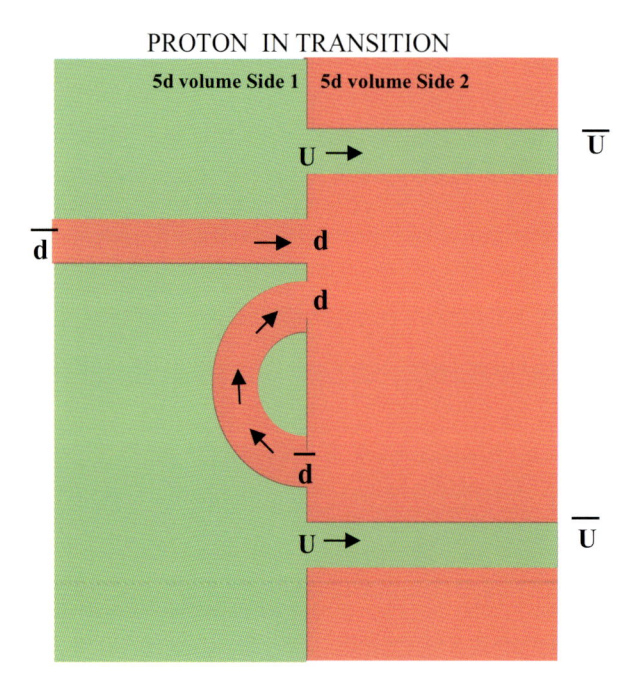

The Down now connects to one of the Ups and the Down creating the Neutron; the Anti-down connects to the other Up quark and creates a Tunneling Positive Pion that moves away from the Neutron in fourth dimensional space where it is invisible.

In the figure below, two particles are created: the **Neutron** and the **Tunneling Positive Pion**.

Figure 36.12

Creation of the Neutron and the Tunneling Positive Pion

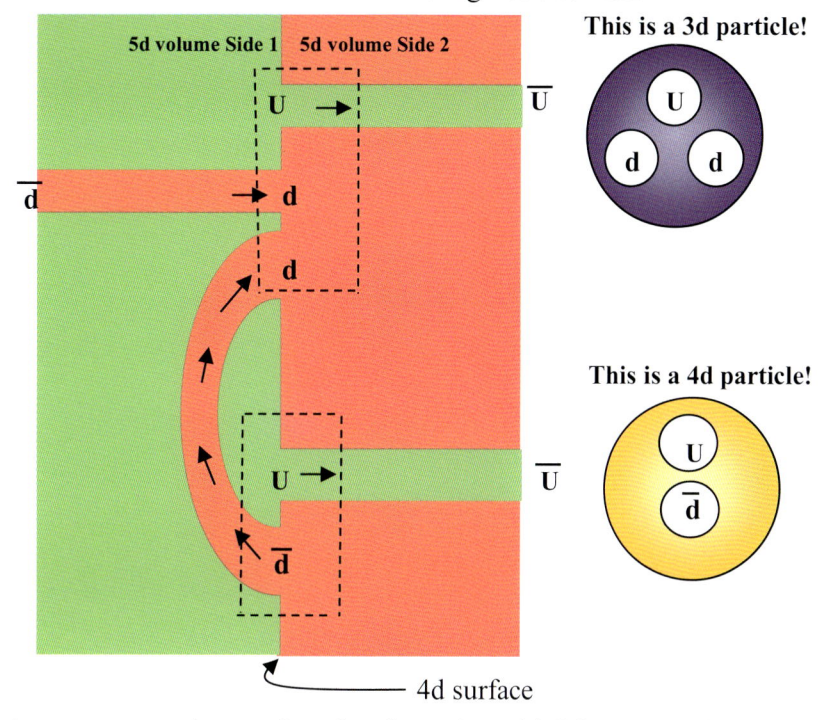

Again it is important to reiterate that the above is a 4d SCHEMATIC DRAWING! It was impossible to simultaneously draw the breaking of the 4d vortex in Figures 36.8→36.11 that creates the positron.

This second, lower dimensional drawing below reveals what happens to the ends of the 4d vortices and their subsequent eruption into 3d space. Note how the vortex creating the Neutron is now twisted into a 4d torus where the entrapped 3d space circles around and around; while the other broken end of the 4d vortex becomes the much smaller diameter 3d hole known as the Positron that contains no quarks. Note how the surrounding 3d space is now flowing into the positron creating its electrostatic charge.

Figure 36.13 The creation of the positron

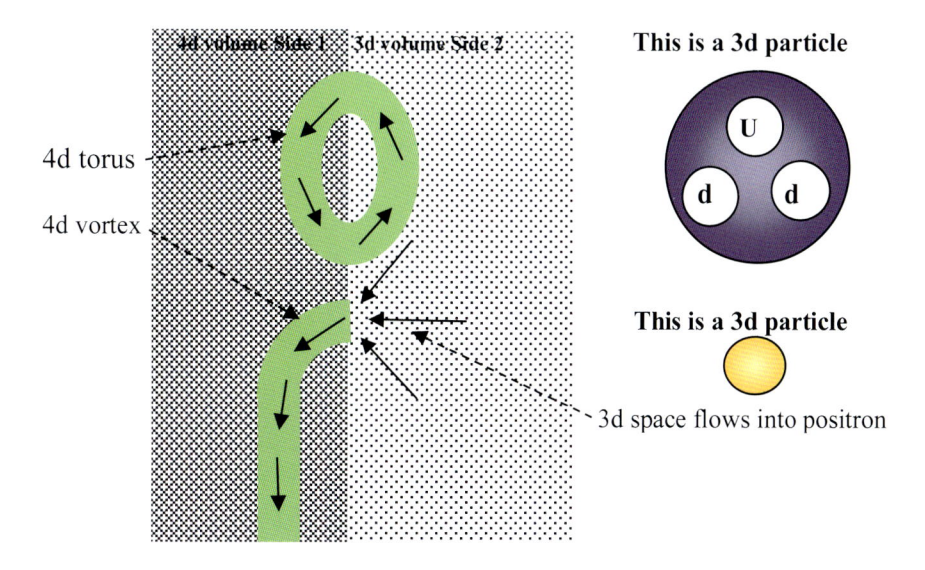

> One of the great achievements of the Vortex Theory is the unification of Quantum mechanics and Newtonian physics. This hundred year controversy can now be put to rest using the discovery of Nuclear Gravity.

Chapter 37
The Unification of Quantum Mechanics And Newtonian Physics via Nuclear Gravity!

One of the great problems with 20th century science is the conflict between Newtonian Physics and quantum mechanics.

Newtonian physics consists of Newton's three laws and gravity; and are successful in describing the motions of the stars, planets, and Galaxies; but they cannot explain what is happening in the subatomic world. Likewise, Quantum mechanics can explain much of the phenomenon of the subatomic world, but it cannot explain Newtonian physics. However, we can now rectify this problem.

The unification of quantum mechanics and Newtonian physics can now be easily made using the discoveries presented in Book 3.

NUCLEAR GRAVITY

Nuclear Gravity is now finally explained by using the principle of less dense space. The less dense space surrounding protons and neutrons in the micro-world is responsible for the force of gravity in the macro-world. In the micro-world of subatomic space, this force is almost negated by the presence of the strong force, weak force, and the electromagnetic force. However, in the macro-world of planets, stars, and Galaxies, the addition of all the less dense regions of space surrounding protons and neutrons – minus the minor denser regions of space surrounding electrons – make gravity the predominate force.

Figure 37.1 "NUCLEAR GRAVITY"

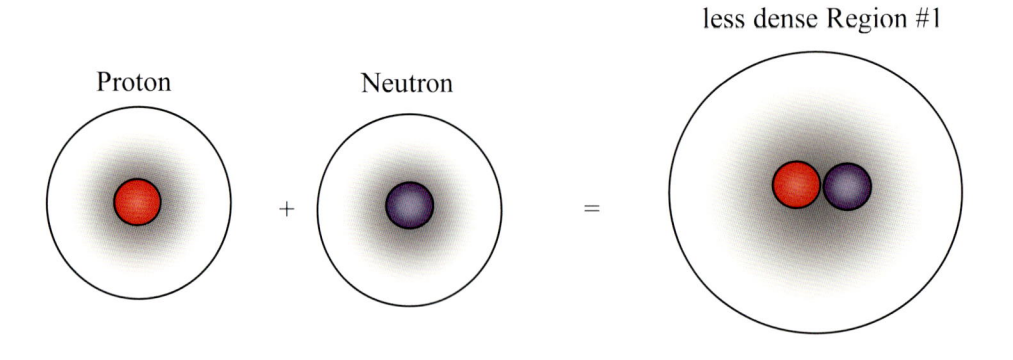

Figure 37.2

Note: region #2 has a slightly lesser volume than region #1. Note too: not to scale.

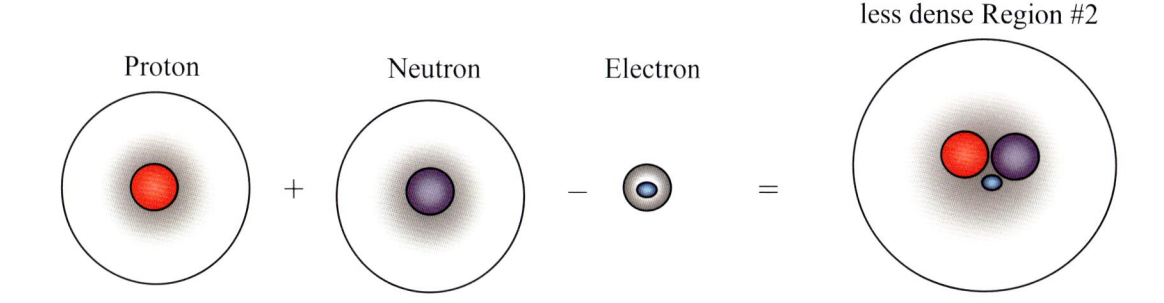

       Proton   Neutron   Electron   less dense Region #2

  Although difficult to draw, the denser space surrounding the electron subtracts from the less dense space surrounding the protons and the neutrons, making less dense region of #1, larger than less dense region of #2. However, even though the electron's denser region of space initially subtracts from the less dense regions of space, when this mass encounters another mass a strange situation develops: the surface of the electron is bent in the direction of the other mass just as are the surfaces of the protons and neutrons! Thus making the electron part of the mass of the gravitation attraction. Because the same situation also occurs with the electrons in the other mass, all the masses of the electrons in both objects attract the masses of the other objects contributing to the force of gravity!

  Consequently, the additions of the hundreds of trillions of the less dense regions of space surrounding protons and neutrons, minus the much smaller denser regions of space surrounding the electrons in planets and stars creates their gravitational field; a field that looks like bent space, but in reality, is a region of less dense space.

Figure 37.3
A Star or planet with its surrounding region of less dense space: creating its huge gravitational field.

Not to scale!

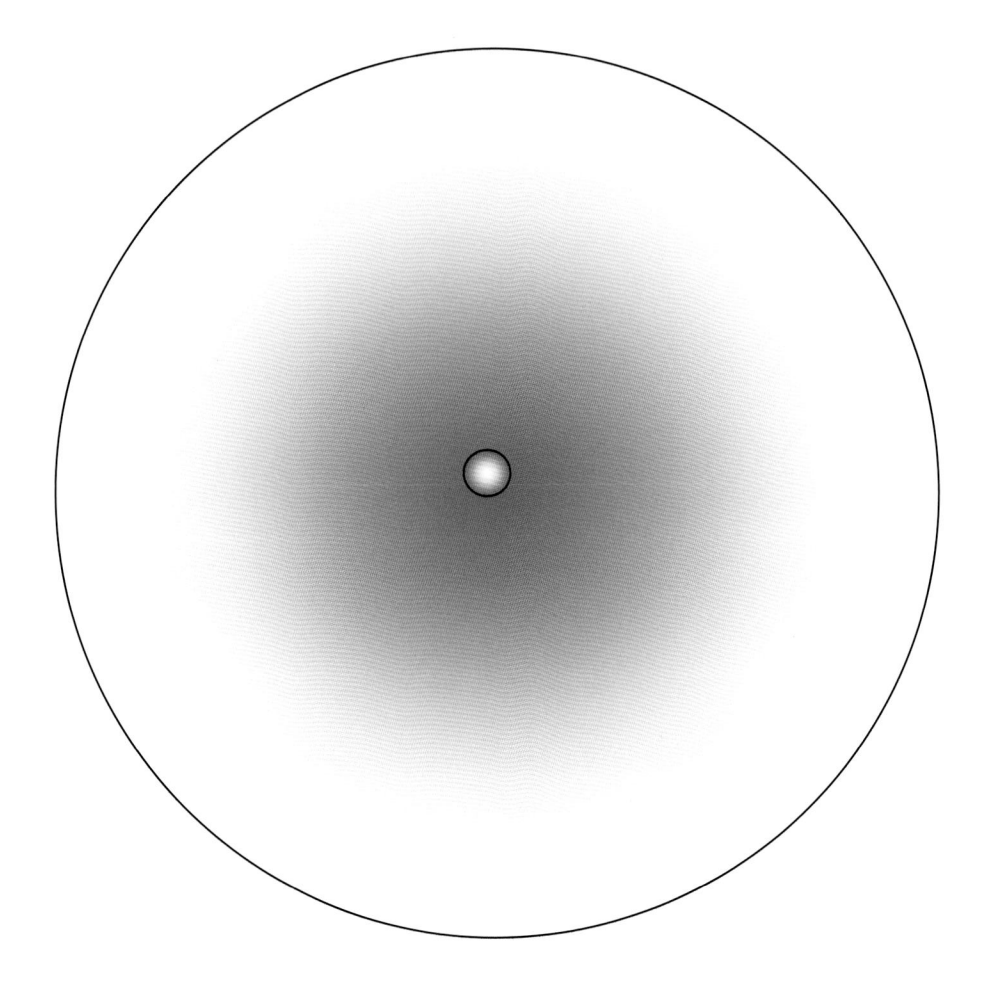

> We now know how the universe is constructed; the question is: "<u>what has it constructed</u>"? One of the most important questions of all is consciousness, awareness, and emotion? If we are nothing but a collection of lifeless holes in space, how is it that our collection has gained awareness? How has the lifeless matter of the universe gained the ability to become alive and be aware of itself? Is this merely a phenomenon created by a certain configuration of the holes in space, or is it a clue to something much, much deeper??? [Some of the following is Paraphrased from Book 1]

Chapter 38
Summing It All Up!

EVERYTHING IS ONE!

The true vision of the universe is the greatest discovery ever made. It is a shocker. A blockbuster:

Stated simply, there are no separate "parts" to the universe.

There is only one gigantic particle; the material it is made out of; and its motion. And that's it! [Shades of Occam's Razor?]

The only thing that exists in the physical universe is the substance of which space is made. Space is not "space". It is not a void with nothing in it". Space is made of something totally unique from our experience. It is constructed out of at least seven dimensions, it can both bend and flow, expand and contract, and have holes in its surface.

This substance is in motion. It is expanding outward at an incredible speed carrying the Galaxies with it: creating their red shift. Matter is created out of space. The "particles" of matter - protons, electrons, and neutrons are not particles at all. They are three dimensional holes existing upon the three dimensional surface of fourth dimensional space. The substance of which space is made flows into and out of these holes, creating their electrostatic charges; their rotations rotate the surrounding space, creating their magnetic effects. The hole creates the particle effect, while the denser or less dense space surrounding the hole creates its wave effect.

Energy is also created out of dense and flowing space. Energy in the form of photons, are really compact, denser regions of space that expand and contract as they move. This dense region - or photon - creates a particle effect, while the expansion and contraction of the surrounding space it passes through creates its wave effect.

Time does not exist; the phenomenon of time is created throughout the universe by the uniform flow of microscopic space into and out of the three dimensional holes of matter. This flow is uniform throughout the universe because it is created by the expansion of space itself which is uniform throughout the universe. [Note: the acceleration of the galaxies is not caused by the expansion of the universe; but rather is a surface effect created by the presence of anti-hydrogen causing nearby galaxies to accelerate away from each other. Both phenomenon, the expansion of the universe and the local acceleration of some of its galaxies are being created by two completely different causes.]

The "forces" of nature are not part of one universal force. The forces of nature are created by bent and flowing space and the exchange of higher dimensional holes as in the case of the strong force.

Everything that exists, every physical relationship, even the creation of the universe itself can be explained by this one simple vision. Hence, the ultimate vision of the universe is also...

THE COMPLETE THEORY OF EVERYTHING...&...THE ROSETTA STONE OF SCIENCE!

This vision also possesses a delicate elegance unique unto itself. Its wonderful simplicity surrounds it with an aura of beauty, unlike anything ever seen before.

To grasp the incredible implications of this vision, first look at or visualize all of the different colors you can see or think of. Look at the reds, the blues, the yellows, the greens, and all of the myriad numbers of colors in-between, and then realize they are all made out of just one silvery "color" possessing different vibratory rates.

Next, visually imagine anything and everything you can think of. Think of the forests, the mountains, the deserts, the oceans, the white clouds in the deep blue sky, or the twinkling stars in the dark night sky and realize this incredible diversity of shapes, sizes and textures are all made out of just one substance. It is a humbling and profound experience.

But there is even a more profound experience: look around the room you are sitting in. Look at all of the different and separate things you see including *yourself.* Now look again at these separate objects and realize *these are not separate objects at all.*

> *EVERYTHING YOU SEE, INCLUDING YOURSELF, IS ONE THING. THERE ARE NO SEPARATE PARTS.*
>
> *SEPARATION IS AN ILLUSION. IT IS AN ILLUSION CREATED BY THE FACT THAT WE CANNOT SEE THE MATERIAL SPACE IS MADE OF. WE ONLY SEE THE HOLES AND NOT THE MATERIAL BETWEEN THE HOLES.*
>
> *EVERYTHING THAT EXISTS EVERYWHERE WITHIN THE PHYSICAL UNIVERSE, INCLUDING EVERY ONE OF US, IS MADE OF JUST ONE SUBSTANCE.*
>
> *WE ARE ALL ONE WITH IT. WE ALL MOVE WITHIN IT. WE ALL EXIST WITHIN IT.*

This vision of the universe is also peaceful. The amazing complexity of the physical universe being created by something so incredibly simple is comforting and reassuring. It is like the deep blue waters of a high mountain lake. In fact, the more you think about it the calmer you will become. Try it yourself.

The next time you are stuck in rush hour traffic on the freeway, imagine the peacefulness and the tranquility of the substance of which everything you see around you is imbedded within. Imagine its quietness, its calmness, even as the chaos of matter seethes *within it.*

When I first did it, it was as if I was at one with the universe; it was meditation; it was the OM: it was a religious experience. A most profound religious experience!

AND THEN THERE IS US!

However, there is something much more important to us all → US!

It is true; scientifically speaking, we are nothing more than a collection of holes in space! And yet we possess unique characteristics: we possess awareness. We are in fact matter; looking back at itself! We also possess intelligence. And we possess emotion. So where did these come from?

Some atheists state that you can program a computer to do all of this! Programmers can mimic awareness, but they cannot program awareness. This is a characteristic unique unto life and only life! Whatever they try to do with computers, they cannot program life into them. Only God can create life! Quantum Mechanics Proponents would have us believe that the universe was created by probability and random acts of matter. But did the fantastically complex DNA molecule just happen to be created by the laws of probability caused by atoms randomly colliding with other atoms as some would have us believe? Look at its fantastic complexity? Are we somehow supposed to believe the DNA molecule just "happened" through accidental chance?

And look at what happens when cells divide. High magnification reveals that the DNA molecule begins to unravel, creating a duplicate of itself which is passed on to the other cell! Is this fantastic ability something that was also just created by the random collisions of atoms? I don't think so!

THE SOUL?

There is something much more to all of us than what our instruments are capable of detecting. So where did this other part of us come from? Its ability to move out of the body and its ability to hover above the body reveal it is not acting like the typical three dimensional matter of this universe. I have my own theories about the construction of the soul, but they are just speculation and personal at the moment. I will leave this question for a future generation of new and revolutionary Religious-Scientists. Hopefully they will take that necessary "one step beyond", and guide mankind into a bright new future.

But getting back to the subject of awareness, it is one thing to fake consciousness in computers by clever programmers programming computers to respond to thousands of different questions. But awareness cannot be faked! And herein is the problem…for where did this awareness come from?

I must confess that I was raised as an atheist but have seen too much in my life to remain an atheist. As explained in Book 1 of this now six book series…reading about a first ever reported near death experience in 1974 changed my life forever. The implications of the possible existence of a soul were too overwhelming to dismiss. I had to know, I had to know the answer if it took my entire life to do it. And that is what has apparently happened.

Although I have grown old in the struggle, I feel good because I have spent my life in a worthy cause. Using the Vortex Theory of Atomic Particles, all of the great mysteries of science have now been solved except for the last one that is the most important of all: God! Is the gigantic particle that everything is created out of, God? We will look at this possibility shortly. But before we do, the following sequence of events must be presented…

LOCATION OF THE KINGDOM OF SOULS & 4D SPACE!

Although it might be hard to believe, it is nevertheless true. Everything that has been discovered in these six books was discovered because of one quote by Jesus in the New Testament. As stated in Book 1, this investigation into the mistakes of science, this quest, began after reading and studying all of the religious texts of the world. I realized that all of religion would be a lie unless a Kingdom for Souls existed somewhere in the universe. Furthermore it had to be a real and not an imaginary place possessing real x, y, & z co-ordinates. So I decided to go looking for this location of the Kingdom of Souls by taking math and science courses in college.

And as reported before, in 1976, while taking a course called Space-time Physics at California State University in Humboldt California I was introduced to the possibility of higher dimensions of space, specifically a fourth dimension of space: an invisible volume of space located at 90 degree angles to our three dimensional space… …possessing *x, y, z, & w* co-ordinates!!! (The *w* makes it real → yet invisible!!!)

Furthermore, movement into or out of this dimension would be called "Without", if you were coming out of it; or "Within", if you were going into it. For me, this created a stunning, shocking revelation, for I suddenly remembered a quote made by Jesus in the New Testament of the Bible…

> *And when he was demanded of the Pharisees, when the kingdom of God should come, he answered and said,* **"THE KINGDOM OF GOD COMETH NOT WITH OBSERVATION: NEITHER SHALL THEY SAY, LO HERE OR LO THERE. FOR BEHOLD. THE KINGDOM OF GOD IS <u>WITHIN</u> YOU."** *LUKE 17:20-21*

And again, as said in Book 1, as you read this quote, try to imagine you are seeing it for the first time. I know for many of you who are Christians it is difficult to ignore the training of a lifetime, but don't cheat yourself. Like everyone else who is not a Christian, you also have a right to know the truth.

Although most Christian peoples are trained from early childhood to believe this statement is a parable or metaphor referring to the heart or mind, is it just a coincidence it is also an exact description of the location of higher dimensional space? By using physical matter (the physical body) as a reference point, the direction of higher dimensional space (if it exists), is "within", towards the very center of matter itself.

It is also important to understand that if Jesus used another reference point it might cause a conflict. For example, if the physical matter of the Earth, Sun, or Moon was used as a reference, it might create a misunderstanding and lead the listener to believe the Kingdom of God was located within the center of one of these celestial bodies. Or, if he had held up a stone and said the Kingdom of God was within, people might believe the kingdom was within that very piece of rock.

As explained in Book 2, this word "within" is also a very important word. Even though other words are adequate such as "inward", or "inside", the word "within" does an excellent job of representing both *the direction and the location* of higher dimensional space. However, as before, the three most important words are "if it exists". For in the present view of the universe then, [45 years ago], with only three dimensions of space and one dimension of time [now proven to be a mistake], there was just no room for higher dimensional space. Furthermore, ideas and speculations about the existence of higher dimensional space are legacies of the modern era of mathematics. During the time of Jesus, higher dimensional space was unknown. But maybe not to Jesus!

When Luke 17:20-21 is analyzed from the point of view of being a direct answer to a direct question, its implications are too intriguing to ignore. The previous misunderstanding of this quote being a parable referring to the heart or mind is easily understood. Unless one knows something about the geometry of higher dimensional space [in Book 1 & 2], this statement appears to be some sort of philosophical reference being made to thoughts or emotions. But this is not so.

To correctly analyze this statement, we must first realize this quote is a direct answer being made to a direct question. Consequently, it is not a philosophical reference that is being made to the heart or the mind, or as to how one thinks or feels. Nor is this statement a simile or a metaphor.

To correctly understand this statement, it must be remembered the Pharisee's believed the Kingdom of God would be an actual kingdom located upon the Earth. A belief contrary to the teachings of Jesus as demonstrated in John 18:36 **"MY KINGDOM IS NOT OF THIS EARTH"**. Consequently, in response to the Pharisee's question, the first words Jesus responds with, **"THE KINGDOM OF GOD COMETH NOT WITH OBSERVATION,"** are referring to this erroneous belief held by the Pharisees, and Jesus' purpose in answering this question is to correct this error.

Jesus then goes on to say, **"NEITHER SHALL THEY SAY, LO HERE, OR LO THERE."** In other words, the kingdom of God *cannot be seen with, or discovered with one's physical eyesight.* (Note: the word **"LO"** denotes exclamations of discovery.) Finally, he ends this statement by boldly stating, **"FOR BEHOLD, THE KINGDOM OF GOD IS WITHIN YOU."** Although these two statements are short and simple, they are most profound. For if the Kingdom of God cannot be seen with one's physical eyesight, it does not exist within our three dimensional universe. Which means the only place it could exist would be in another dimension of the cosmos.

Because all dimensions are at right angles to each other, the only way to enter the fourth dimension when using matter as a reference point is to travel at right angles to all three dimensions simultaneously, or "within" - towards the very center of matter itself [See the explanation of 4d space in Book 1 or 2]. The exact direction Jesus is alluding to when he states, **"FOR BEHOLD, THE KINGDOM OF GOD IS WITHIN YOU."** He adds special emphasis to this statement by adding the word, **"BEHOLD"**: which means - "look", "see". A word he seldom used except to emphasize an extraordinary statement - a revelation.

Here we must pause and ask ourselves if we are expecting too much of Jesus. The concepts of higher dimensional space belong to Twentieth Century cosmology, not to ancient religious philosophy.

Or do they?

If Jesus was answering an erroneous question with an accurate view of reality, he already knew about higher dimensional space and didn't have to wait for modern man to invent it.

If this analysis is indeed correct, if it is in fact a correct explanation of Luke 17:20-21, its implications leave us stunned. For if the Kingdom of God does indeed exist in higher dimensional space, it means higher dimensional space exists, and if it exists, the present vision of the universe with its three dimensions of space and one dimension of time is in error. Hence, it means that if by using the words of Jesus in the New Testament we can find that error, correct it, and discover the true vision, the course of human history will be altered. [This error has now been identified and corrected via *The End of Time* PhD thesis.]

Although Christian peoples already believe Jesus as the Messiah "knew things no man has ever known", for the first time ever, the rest of the peoples of the world will know it too. They will have scientific proof Jesus' knowledge of the construction of the universe exceeded and surpassed the knowledge discovered by all of the greatest scientists who have ever lived.

It must also be emphasized that the knowledge of the existence of higher dimensional space is not just any knowledge. It is very difficult - extremely difficult knowledge to obtain. This knowledge stands upon the apex of a pyramid of other scientific discoveries. It is knowledge obtained only by the most precise and exact experimentation.

Even more important, it soon became apparent higher dimensional space is not just some mundane, insignificant place. Quite the contrary! Higher dimensional space is the most important place in the cosmos. It was discovered that without higher dimensional space, nothing can exist within our three dimensional universe. Without higher dimensional space, there are no fourth dimensional vortices. And without them, no three dimensional holes; no matter; and no us!

In knowing all of this, can there then be any doubt Jesus was the Messiah? Will the attention of the people of the world shift towards this most wonderful man of all men? Can a renaissance in Christianity be created? And as asked in Book 1…the stakes are high. Can it be done?

Can the course of human history be altered by one quote from Jesus?

Yes it can, and it will! The error in the scientific vision of the universe was found. It was time, "Time itself"! It does not exist, causing us to re-evaluate the entire scientific vision of the universe: which was done in the previous six books. But an even greater, more stunning conclusion now exists: Have we discovered God!

HAVE WE DISCOVERED GOD??

What have we accidentally discovered?

A great philosopher once said, "Know thyself".

I guess I never knew myself until I asked myself the question, "Is this God?" because when I did, I was suddenly afraid.

Just what had I stumbled onto?

I was just a mental adventurer in search of the thoughts no man had ever thought before. But never, never in my wildest dreams, did I ever even think I might discover God.

Had I?

How can such a question even be answered? How *would* anyone know if they had discovered God? How *could* anyone know if they discovered God?

Is there any way to recognize God? How can God be identified? What is the definition of God?

Looking in the Bible for a description of God we see terms such as the "*Alpha and the Omega*", and the "*I am, That I am*".

Using these terms, it is easy to see that the substance out of which space is made is certainly the Alpha and the Omega, the beginning and the end of everything that exists in the physical universe, but is it the "I am, that I am" of the Bible?

Just what is it? Is there a way to find out?

As reported before: scared of what I had found, I was grateful for the opportunity to turn my mind to another subject. I looked for a way to resolve this dilemma. And while thinking about it, I realized the answer could be found in the creation of the universe.

I realized that if God created the universe, there are three ways he could have done it: the universe is created out of God; the universe is a unique manifestation of God; or the universe itself is a creation of God.

If the universe is created out of God, the space of the universe is God, and *we exist* within the actual physical body of God. This vision is comparable to the existence of a single biological cell living within our physical body. (A single microscopic creature existing within a much larger, massive creature.)

Or, if the universe is a unique manifestation of God, the space of the universe might be a thought, and we might be mental projections within this thought form. Although this idea might seem farfetched, we must never forget that we dream. In our dreams we appear to see real images, and have real experiences. [How does this happen, how are these images formed, and why do they appear to be so real?]

But if the universe itself is a creation of God, then space is a creation of God, and we are creations within the creation. This scenario can be compared to a man inventing a liquid plastic, forming bubbles within its volume, and then somehow arranging the bubbles to form images.

So which is which?

Another question: can a mere mortal man even answer such a question? Are there limits to the mental constructs our mind is capable of forming? The mental picture of God being indescribable and incomprehensible to the mind of a mere man is one of the accepted beliefs of world religion practitioners. But is it true? Can a creation analyze its creator? Is it a sin to even try?

But then again, maybe it would be a sin not to try! Maybe I was reacting to the inherited fears of our ancient ancestors. The same fears that kept them from crossing the ocean because they were all afraid they were going to fall off of the edge of the earth. Or the fears that kept them from challenging the hypocritical doctrine of the church leaders, and allowing the inquisition to take place because they believed these maniacal priests were doing the will of God.

In the past, men also believed in demons. They tortured, and murdered other men in fear of them. However, they failed to realize the only real demons were the fear and ignorance within their own minds. Fear and ignorance; ignorance and fear! The true rulers of men! Allowed to run rampant and without restraint they accentuate each other, intensify each other, until they dominate the minds of men.

But no matter how powerful they may seem to be, they are only thoughts. Because they are only thoughts, the way to beat them is with another thought or with no thought at all: to totally ignore their presence. And that is exactly what I did. I told myself that my fears were mental projections put into my mind by ignorant men and gave them no more credence. Then, I boldly returned to the subject of the moment: is the space of the universe God?

While thinking about it, I began to recall all of the religious texts of the world. I remembered that wherever men believed in God there seemed to be three words everyone everywhere in the world agreed upon. Three words universally accepted and associated with God in all religious beliefs. These words are *omnipresent, omniscient, and omnipotent*. They speak for themselves:

If the substance out of which space is made is God, then God is surely *omnipresent*. God is everywhere. Every "particle" of matter and every photon of energy everywhere in the physical universe is made out of him. Since he would be aware of every motion of everything that exists everywhere in the universe, he would know "…when a single sparrow falls from the sky…". He would also know the configurations and the interconnections of every neuron within a man's mind. He would know when electrical current flows between them, and in doing so, he would know every thought a man thinks.

Knowing every thought a man thinks, there would be no fooling God. He would know the hypocrisy, the lies, and the true feelings of every man in the world. Knowing all and seeing all, he would know the secrets of every person everywhere in the universe.

He would know when a man earnestly prayed to be heard by him, or to really be seen by other men. He would know what happens in the light of day or in the darkest dungeon in the world. His knowledge would be all encompassing, and universal. He would indeed be *omniscient*.

However, is he *omnipotent?*

It is easy to see how the substance out of which space is made is omnipresent, and omniscient, but to be omnipotent is another matter entirely. The deep contemplation of this problem reveals that the key to its answer is found in motion. The creation of the universe began when space was put into motion. But did space set *itself* into motion, or did something else do it?

[A very religious friend of mine, reading what I had written said this, "Remember what it says in Genesis 1:2…and the spirit of God <u>moved</u> upon the face of the waters…"!]

If something else set it into motion, it is not omnipotent and it is not God. However, if the substance out of which space is made set itself into motion, then *it* created the universe out of itself. It made everything, and in doing so, became omnipotent.

To be omnipotent, means it had the ability to create the universe and used that ability. This shows premeditation and cognizance of thought. Which means space had to have the mental ability to both think and reason before the physical universe was created. However, all of these ideas are predicated upon the premise that space possesses awareness, consciousness. Does it?

While thinking of a way to answer this question, surprisingly, the following answer seemed to just pop into my mind all by itself. It begins like this: where does our awareness - our consciousness - come from? In fact, where does any animal's awareness come from?

If we could take a human being apart, atom by atom, we would find we would have several thousand piles of compounds. If we could then break these compounds apart and separate them into the individual elements they are constructed out of, we would have about a hundred piles of individual elements. And finally, if we could take these individual elements apart, we would end up with just three big piles of electrons, protons, and neutrons.

But these are inanimate objects. They possess no consciousness or awareness. Furthermore, all three of these piles are not "matter" at all. They are just three massive collections of three dimensional holes in space. So we are left with a stunning possibility: does a particular arrangement of three dimensional holes create consciousness?

And something even more fantastic that is a fact: the inanimate objects of the universe have obtained consciousness of themselves. When we look at a proton or an electron, in reality, *it is the inanimate matter of the universe that is looking back at itself*!!!!!!!!!!!!!!! A simply astounding idea!

And it is this astounding idea that leads us to another even more astonishing idea: if a creation made out of nothing but holes in space can obtain consciousness of itself, can the space it is created out of obtain consciousness of itself too?

If it can, and if it set itself into motion creating the physical universe - and us in the process - then my friends, it is God.

> *And suddenly, for the first time in human history, we see possible scientific evidence that God actually exists! Evidence that cannot be ignored!*

FOR LAWYERS?

When the ramifications of the Vortex Theory of Atomic Particles were all worked out, shockingly, it was realized that the existence of God is now a real possibility! For everyone who believes in God, this stunning revelation of science is one of its greatest triumphs. The circle is now complete. When science broke away from religion to discover the truths of the universe, one of the truths it discovered was the scientific proof that the existence of God is a real possibility!

Perhaps the ultimate irony of all ironies occurs when the vortex theory is applied to the atheistic viewpoint of the universe. From the atheistic viewpoint of the universe, it now has to be concluded that the existence of God is a real possibility! Here's why: from the atheistic viewpoint of the universe, the consciousness of men is a function of matter. However, the Vortex Theory has discovered that matter does not exist the way it appears to exist.

The protons, electrons, and neutrons that all the matter in the universe is made of are really three dimensional holes existing within the surface of the fourth dimensional substance space is made of. Because of this fact, the physical bodies of men are nothing more than a massive collection of holes existing within this same substance. They do not have a separate existence apart from this substance. Nevertheless, these holes have obtained consciousness: **Awareness!**

Consequently, if a collection of holes within this substance can obtain consciousness, it is also possible that the substance the holes are made of could gain consciousness of itself too! And if it has, and if it set space into motion, creating the physical universe, then it is God!

Because of this logic, the existence of God is now a **real possibility** that must be recognized by science. When presented with this evidence, anyone who does not acknowledge that the possible existence of God is now a real possibility → is acting illogical and irrational! **Atheism is now dead! To become aware of the facts, and continue to be an atheist is to now be recognized as an illogical and irrational person!**

For lawyers: are the beliefs of the irrational to be forced upon those who choose to think rationally? I don't think this is how our system of justice was created to function!

PART VIII
THE APPENDIX

> Who says scientists do not believe in God? Yes, I suppose a few don't, but you will be surprised to find out just who did! The following religious information is never mentioned in the science books or allowed to be used by educators in public schools!
>
> Some of the most famous who were devout believers in God were: Copernicus, Galilei, Boyle, Newton, Kepler, Volta, Faraday, Maxwell, Mendel, Hertz, Pasteur, Joule, Kelvin, Thomson, Carver, Eddington, Millikan, Born, Compton, Einstein, Planck, & Heisenberg.
>
> Note too: According to *100 Years of Nobel Prizes*, a review of Nobel prizes awarded between 1901 and 2000 reveals that (65.4%) of Nobel Prizes winners, have identified Christianity as their religious belief. Christians have won a total of 72.5% of the Nobel Prizes in Chemistry, 65.3% in Physics, and 62% in Medicine.

#1 WHO SAYS SCIENTISTS DO NOT BELIEVE IN GOD???

If you attended or attend public schools in the United States, you will receive a censored education!!!

The public institutions in the U.S. love to tell us how educations in communist countries continually feed their populations propaganda and false histories. However, they neglect to tell the students in all of the public classrooms in the U. S. about the incomplete histories they are forcing teachers to teach students in science classrooms!

Shockingly, most of the most famous scientists who have ever lived believed in God! So why don't public educators want the students at their institutions to know this?

They like to talk about the so called need to keep science and religion separate from each other; but in doing so, completely ignore the fact that many of the discoveries famous scientists made, and even some of the ideas they rejected were motivated by their beliefs in God! Perhaps the most famous example of a rejection was Einstein's denunciation of some of the proposals made about quantum mechanics.

In his response to the proposal of a universe based upon quantum mechanic's insistence upon randomness and probability, Einstein declared, "God does not play dice!" Some public schools even have posters of Einstein and this statement. However, the same schools ignore the logical conclusion that if he did not believe in God, he would never have made such a statement: something they do not want to talk about. Consequently, Einstein's powerful belief in God is ignored by public institutions. So here are some of the beliefs of many of the most famous and notable scientists whose blockbuster discoveries have shaped our lives for the past 500 years…

Nicolaus Copernicus: who can forget that this great man put the Sun at the center of the solar system, and in doing so became famous as the founder of what has since become known as the *Heliocentric Model* of the universe. But what is not known, and is not made famous is the fact that he also served as a Canon of the Church in Frombork, in the region of Polish Warmia where he lived! A church he became a member of because of his belief in God!

Then there is **Galileo Galilei**; who has not heard of him dropping objects off of the Leaning Tower of Pisa and watching them fall to earth. Also, many know about the many statements he made that ended up offending the church "officials" in that era. That is known and is talked about in public schools. When someone speaks against the church, public school teachers love to tell their students about it.

And yet, what the teachers at public schools never talk about is another statement he made! Perhaps his most famous statement is never talked about: this great Italian astronomer, mathematician, physicist, philosopher, and engineer is not quoted in today's public schools as saying, "God is known by nature in his works, and by doctrine in his revealed word!"

Robert Boyle is not as famous as Galileo, nonetheless, *Boyle's Law* which describes the inversely proportional relationship between the Absolute Pressure and Volume of a Gas, has to be learned by every chemistry student in the world. What is not that well known is that he was a devout Anglican and his many other writing's about Theology are never talked about in chemistry classes! Why haven't your children's teachers told them about his beliefs in God?

Another hall of famer is of course the famous **Isaac Newton**. I doubt that there is anyone who has ever studied science that does not instantly recognize his name. Especially since the unit of force, the *Newton* is named in honor of him.

But shockingly, what most people don't know is that on his death bed, Newton confessed to having spent the last years of his life devoted to studying the words of Jesus in the New Testament; and that he never married and remained a virgin all of his life because of his faith and belief in Jesus Christ and God? Why hasn't this ever been told in public High Schools?

Also, there is not one astronomy student who has not heard of **Johannes Kepler**. His three laws of planetary motion, called *Kepler's Laws* are required study, to be learned by all astro-physicists. However what most of these students do not know was that Kepler also incorporated religious reasoning into his scientific works! These religious beliefs were motivated by the belief that God had created the world according to what he called an *"Intelligible plan"!*

Alessandro Volta is next. This Italian physicist invented the first electric battery. The electric unit *Volt* was named in honor of him. In school, every student of physics and electrical engineering knows his last name. What they have not been told in school is his confession:

"I do not understand how anyone can doubt the sincerity and constancy of my attachment to the religion which I profess, the Roman Catholic and Apostolic religion in which I was born and brought up, and of which I have always made confession, externally and internally… but through the special mercy of God I have never, as far as I know, wavered in my faith... In this faith I recognize a pure gift of God, a supernatural grace…May this confession which has been asked from me and which I willingly give, written and subscribed by my own hand, with authority to show it to whomsoever you will, for I am not ashamed of the Gospel, may it produce some good fruit!

An obscure individual most people have never heard of, yet are well acquainted with and aware of his last name is **Andre Marie Ampere**. He is one of the founders of electromagnetism. The unit for electric current that runs every household in the world; the *Ampere* is named after him. He was also a Catholic who was said to always go to the Bible in times of stress or tragedy for solace.

Michael Faraday, was a British scientist who added to the knowledge of electromagnetism. His main discoveries include the principles underlying electromagnetic induction, diamagnetism and electrolysis. It is important to note that Ernest Rutherford's statement about him and quoted in public institutions is this: "When we consider the magnitude and extent of his discoveries and their influence on the progress of science and of industry, there is no honor too great to pay to the memory of Faraday, one of the greatest scientific discoverers of all time."

But he was also a devout Christian. He belonged to a branch of the Church of Scotland known as the Sandemanian Denomination. But more than a member, Faraday served as deacon and elder in this church. Why isn't this spoken of in public schools? Why do science teachers avoid talking about this great scientist's belief in God?

James Clerk Maxwell is another giant of physics and electrical engineering. The mathematics he developed explaining electric and magnetic fields are called *Maxwell's Equations*. These Partial Differential Equations are required knowledge for all students of physics and electrical engineering.

But never spoken of are his deep Evangelical beliefs that he described as, "having given him a new perception of the Love of God"! Where is this spoken of in books of electrical engineering?

Then there is **Gregory Mendel**. Mendel's pea plant experiments conducted between 1856 and 1863 established many of the rules of heredity, now referred to as the laws of *Mendelian Inheritance*. But he also became a friar and was given the name Gregor when he joined the Augustinian friars.

Mendel entered the Augustinian, St Thomas's Abbey in Brno, Margraviate of Moravia and began his training as a priest. In 1867, he became Abbot of the monastery. Why is this not talked about in courses in biology?

Heinrich Hertz proved the existence of electromagnetic waves. In honor of his discovery, the frequency called *Hertz* is named after him. It is a tragedy that he died at 36 years of age. He accomplished much. Research institutions are named in honor of him. The IEEE Heinrich Hertz Medal, was established in 1987, and is awarded "for outstanding achievements in Hertzian waves. It is given each year to a worthy individual for an achievement which is theoretical or experimental in nature". But not mentioned in science or engineering books is the fact that Heinrich Hertz was also a Lutheran: a Lutheran whom the Nazis tried to erase from history because he had some Jewish ancestry! Are public school teachers doing any less in regard to erasing his beliefs by not talking about them?

Next we come to **Louis Pasteur**: world famous for his discovery of Vaccination and Pasteurization. Every country and person in the world owes him a debt of gratitude.

Many science institutes are named after him. He is honored in all public schools in America. The process called *Pasteurization* used to sterilize foods is named in honor of him.

However, how many have heard about his faith? Pasteur remained an ardent Christian throughout his whole life. His step-son wrote: "Absolute faith in God and in Eternity, and a conviction that the power for good given to us in this world will be continued beyond it", were feelings which pervaded his whole life; the virtues of the gospel had ever been present to him. Full of respect for the form of religion which had been that of his forefathers, he came simply to it and naturally for spiritual help in these last weeks of his life".

James Prescott Joule: every student of chemistry and physics in public schools has heard of him. He discovered the law of the conservation of energy, which led to the discovery of the first law of thermodynamics. The SI derived unit of energy, the *Joule*, is named in an honor of him.

But what students of chemistry and physics do not know of him is this written by a biographer: "all of his work was accomplished while he adhered to a belief in an eternal and benevolent God as the creator of the universe. Joule saw the beauty and harmony of nature and its underlying laws as God's handiwork". Who ever heard of this from a teacher teaching chemistry or physics?

Lord Kelvin was a man of many achievements in a number of different fields of endeavor in both science and engineering. He was Knighted by the Queen for his work on the transatlantic telegraph. His name was also given to the *Kelvin Temperature Scale*. A scale that begins at the temperature of absolute zero that all students of science have to memorize: –273C.

What they don't memorize however, because it is never taught, is that Kelvin remained a devout believer in Christianity throughout his life. According to biographers, chapel attendance was part of his daily routine. He saw his Christian faith as *important* to his scientific work. This was imparted to the public during his speech to the Christian Evidence Society, 23 May 1889.

J. J. Thomson! One of the greatest unknown scientists the public has never heard of; yet he must be remembered and never forgotten because he was the discoverer of the electron that has made this era of electric technological miracles possible. He was also a Christian and a member of the Anglican Church. So the next time we turn on an electric light we must thank Mr. Thomson for it! Without him, there would be no cell phones or television!

George Washington Carver was to the biological sciences as Newton was to physics. His teachings of crop rotation and the various uses of peanuts made him a folk hero in the Southern U.S.

Carver also believed in Jesus and combined both religion and science to his teachings at Tuskegee University.

One of the greatest British Astronomers of the 20th Century was **Arthur Eddington**. Actually, Sir Arthur Stanley Eddington is more appropriate because he was Knighted in 1930. All astronomers have heard of him, as well as students of physics. What most of them never heard however was that he was also a devout Quaker and gave his prestigious *Gifford Lectures* at the University of Edinburgh in 1927 on his beliefs in God and science.

Robert Millikan is another one of those great unknown scientists. He is noted for the great achievement of measuring the electric charge of the electron: for which he won a Nobel Prize. He taught at the University of Chicago, and at Cal. Tech. in Southern California.

However, as a lifelong Christian, even more important were his teachings about the importance of both science and religion, as noted in some of his *Terry Lectures* at Yale published as: *Evolution in Science and Religion.*

The German Physicist and Mathematician **Max Born** makes the list because of his efforts to develop Quantum Mechanics. It seems rather odd that this Nobel Prize winning scientist whose endeavors were devoted to the randomness of the universe should be heralded as a man who also believed in God! And yet it's true.

Max Born was Jewish and had to escape from Nazi Germany before World War II!

Then of course there is **Albert Einstein**. Winner of the Nobel Prize, whose picture is plastered in almost every public school in the United States, is known as America's greatest scientist. What is not emphasized is that Einstein also believed in God; so much so in fact that he outright rejected quantum mechanics because of its proposal that the universe was governed by the laws of probability. Something that offended Einstein so much he stated: "God does not play dice!"

Werner Heisenberg: this Nobel Prize winner is famous for the *Heisenberg Uncertainty Principle* that all students of physics have to learn, represents a stark contrast to Max Born. Heisenberg, a German, was also Lutheran. But what is shocking to contemplate, was that he was also a part of the Nazi scientific hierarchy during World War II? So the question becomes, how can a man justify being both a Christian and a Nazi? Or rather, how can a Christian whose message is love; be part of an organization whose destiny and purpose is to murder and subdue humanity? [I don't understand this guy! Did he endorse Nazism to escape persecution?]

He is a direct contrast to our next and most famous, and courageous, of all German Scientists…

…**Max Planck**! Mr. Planck is perhaps the most famous of all German scientists. A Nobel Prize winner for discovering what is now called *Planck's constant*, a number used by every scientist in the world. Equally important, he is also recognized as one of the founders of Quantum Mechanics. But what is not known by many who use his *constant* in their scientific endeavors is Planck possessed great moral courage. This great German was one of few who were bold enough to directly oppose Adolf Hitler. In fact his son was executed in the failed plot to assassinate Hitler. Perhaps this was due to his and his father's belief in God. Here is something that this genus of science and mathematics said about religion and God:

"Religion represents a bond of man to God. It consists in reverent awe before a supernatural Might, to which human life is subordinated and which has in its power our welfare and misery. To remain in permanent contact with this Might and keep it all the time inclined to oneself, is the unending effort and the highest goal of the believing man. Because only in such a way can one feel himself safe before expected and unexpected dangers, which threaten one in his life, and can take part in the highest happiness – inner psychical peace – which can be attained only by means of strong bond to God and unconditional trust to His omnipotence and willingness to help." (Max Planck 1958).

Also…

"…all matter originates and exists only by virtue of a force which brings the particles of an atom to vibration and holds this most minute solar system of the atom together. . . . We must assume behind this force **the existence of a conscious and intelligent Mind. This Mind is the matrix of all matter.**" – Max Planck

Although there are many other great scientists who also believed in God, there is not enough room in this book for them all. Hence we conclude with **Arthur Compton** whose faith in God was unflinchingly stated.

Compton was an American who won the Nobel Prize for discovering the "*Compton Effect*" [later named as an honor to him]. His devotion to science and teaching is well documented. He was both a professor of Physics and a University Administrator. He wrote many important scientific papers on the Compton Effect; and the equally important discovery that the intensity of cosmic rays was correlated with geomagnetic rather than geographic latitude: revealing that they were really charged particles! But perhaps the following statement made by him *not* regarding science is his most important ever…

Commenting on the Bible in the Chicago Daily News (April 12, 1936), Arthur Compton said this: "For myself, faith begins with the realization that a supreme intelligence brought the universe into being and created man. It is not difficult for me to have this faith, for it is incontrovertible that where there is a plan there is intelligence. An orderly, unfolding universe testifies to the truth of the most majestic statement ever uttered: 'In the beginning God…' [Genesis 1, 1]." (Compton 1936).

The above scientists were singled out of the lists of Christians and Jewish practitioners not only because of their fame, but because of the fact that scientists and engineers use the *scientific units* named after them every day in their research and engineering projects. Without them, there would be no science or engineering projects!

#2 LIST OF REVOLUTIONARY DISCOVERIES

Unification of Newtonian Physics and Quantum Mechanics
Discovery of the 5th force in nature: the Anti- Gravity Force!
The Force of Gravity Explained
The secret of Quantum Entanglement explained
The Explanation of how single photons create Double Slit Interference Patterns
The Explanation of The Pauli Exclusion Principle
The Explanation of Dark Energy
The Explanation of Dark Matter
The explanation of "Tunneling"
Dispelling the Myth of the Higgs Boson Particle
The First ever Explanation of the Constant of Fine Structure: the 1/137 Mystical dimension-less number all of the greatest scientists of the past 100 years have tried to explain!!!
Explanation of the Asymmetric Parity of Neutrinos
How the Particle and Wave Theory of Light is created
The Explanation of the Striking Parallel Between Newton's Law of Gravity and Coulomb's Law
An Explanation of the Creation of the Universe [unlike anything ever proposed before]
An Explanation for how the Universe will end [unlike anything ever proposed before]
The Explanation for the creation of the phenomenon of Time
The Explanation for Time Dilation Effects at near Light Velocities
The Explanation for length shrinkage at near Light Velocities
Explaining the Mystery of Mass [no Higgs Boson is needed!]
The Electromagnetic Force Explained [like never before]
The Weak Force Explained [like never before]
The Strong Force Explained [like never before]
The Anti-gravity Force is introduced and Explained
How the Particle and Wave Theory of Matter is created
The Explanation of Intrinsic Spin [1/2 spin]
The Explanation of Newton's Three Laws of Motion
Resolving the Conflict between Inertial Mass and Gravitational Mass
The Explanation of the Conservation of Charge
The Explanation of the Conservation of Angular Momentum
Explaining the Conservation of Momentum
The Conservation of Mass and Energy is explained
The Explanation of the Mystery of Entropy
What the Neutrino really is!
The true Explanation of Buoyancy
The Explanation of how the Proton is created

The Explanation of how the electron is created
The explanation of how the Neutron is created
The Explanation of Covalent bond in Chemistry
The explanation of the Ionic bonds in Chemistry
The Mechanical Explanation of the Michelson Morley Experiment
The <u>Five</u> Forces in Nature are explained using configurations of space!
The Explanation of Mass
The Explanation of Energy
What causes Acceleration
The Explanation of the Muon's Prolonged Lifetime when moving at Relativistic Velocities
All Phenomenon Associated with the Theory of Relativity are now Explained
The explanation of the Strange Perihelion progression for the planet Mercury
Why all Particles possess the Same Amount of Charge
The Explanation of Black Holes
The Explanation of Planck's Constant
The Explanation of Increasing Velocity and Increasing Mass
The Grand Unification Theory
Why Electrons Orbit Protons
The True Vision of Space
How the Proton and the Electron create a hydrogen atom with two flowing vortices
The Vortices in higher dimensional space
Particle Collisions
The True Vision of Energy
The end of Einstein's Spacetime!
The End of the Theory of Relativity
Creation of the ±1 Charge,
The creation of Spin
Creation of the Up and Down Quarks
Creation of the Strange and Charm Quarks
Creation of the Top and bottom Quarks
"The Four Layers of Matter"
Leptons Explained
Neutrinos Explained
How Particle Collisions Create New Particles
The Explanation of how Quarks Change "Flavor"
The Explanation of the Law of the Conservation of Lepton Number
Lepton Creation during the Decay of Positive and Negative Pions
Neutrino Creation during the Decay of the positive Muon
Neutrino Creation during the Decay of the Negative Muon
The Decay of the Positive Muon
The Decay of the Negative Muon
The Collision between a Proton and an Electron Anti-neutrino

The collision between a proton and a Muon Anti-neutrino
The Collision between a Neutron and an Electron Neutrino
The Collision between the Neutron and the Muon Neutrino Collision
The Decay of the Neutron and the Creation of the Anti-neutrino
The Explanation of the Law of the Conservation of Baryons
The Explanation of the Conservation of "Strangeness"
Gauge Bosons are not Force Carriers between Particles
The Explanation of the CPT Theorem
The creation of the Pentaquark and the "Neutral Pentaquark"
The Motion of Photons and Particles through Electric and Magnetic Fields
The Reason why the Photon's Electric and Magnetic Fields exist; and why they are at Right Angles to each other
The Stability of Protons; the Instability of Mesons
The Explanation of what Quarks are
The Explanation of Quark Confinement
The Two Sides of Space!
The Two volumes of Space: one increasing; one decreasing
The REAL Explanation of the 1/3 & 2/3 Charges of Quarks
The Explanation of ±2 Charge of Resonances
The Explanation of 3/2 spin
The Explanation of the Up Quark
The Explanation of the Down Quark
The Explanation of the Strange Quark
The Explanation of the Charm Quark
The Explanation of the Bottom Quark
The Explanation of the Top Quark
The Explanation of the Muon
The Explanation of the Tau
The Explanation of the Electron, Muon, and Tau Neutrinos
The Difference between Strong Force and Weak Force Creations
The Explanation of how Quarks Decay into other Types of Quarks
The Explanation of "The Law of the Conservation of Lepton Number"
The Explanation of Neutrino Creation during Neutron Decay.
The Explanation of the "Law" of the Conservation of Strangeness
The Explanation of the Strange Quark's Extremely Long Lifetime
The Explanation of the W Particle
The Explanation of the Z Particle
The Explanation of the CPT Theorem
The Explanation of the Pentaquark.
The Proposal of a new Particle in Nature: THE NEUTRAL PENTAQUARK
Dispelling the Myth of Gluons
Dispelling the Myth of Gravitons
Quantum numbers are effects and not causes! [Explain the Pauli exclusion principle.]

Dispelling the Myth of the Quantum Field Theory
[Note: how can you have a field that supposedly permeates the entire universe yet violates the very principle of a field: no point of origin; does not dissipate with the square of the distance.]
The Explanation of Anti-matter and Dirac's mistaken explanation
Creation of the Alpha particle
How the nucleus of an atom <u>creates</u> gamma rays and x-rays
How fusion creates energy
GOD! Has GOD been discovered!!!
Universal Religion?

References

National/International Conferences attended, and peer reviewed scientific papers presented

[1] The Vortex Theory of Matter. [Presentation of his own work]
'International Forum on New Science' Colorado State University (1992, Sept 17-20).
Moon. R. Fort Collins, Colorado. USA. Topic: The Vortex Theory of Matter. Copyright 1990)

[2] The Vortex Theory and some interactions in Nuclear Physics. [Book of abstracts; p. 259]
'The LIV International Meeting on Nuclear Spectroscopy and Nuclear Structure; Nucleus 2004'
(2004, June 22-25). Moon, R., Vasiliev, V. Belgorod, Russia.
http://nuclpc1.phys.spbu.ru/nucl/Abstracts/Nucleus_2004.pdf

[3] The Possible Existence of a new particle: The Neutral Pentaquark. [Book of materials; pp. 98-104]
'Scientific Seminar of Ecology and Space' (2005, February 22). Scientific Research Centre for
Ecological Safety of the Russian Academy of Sciences. Moon, R. Saint Petersburg, Russia.
https://spcras.ru/ensrcesras/

[4] Explanation of Conservation of Lepton Number. [Book of materials; p. 105]
'Scientific Seminar of Ecology and Space' (2005, February 22). Scientific Research Centre for
Ecological Safety of the Russian Academy of Sciences: Moon, R., Vasiliev, V. Saint Petersburg,
Russia.
https://spcras.ru/en/srcesras/

[5] Explanation of Conservation of Lepton Number. [Book of abstracts; p. 347]
'LV National Conference on Nuclear Physics' (2005, June 28-July 1). FRONTIERS IN THE
PHYSICS OF NUCLEUS. Moon, R., Vasiliev, V. Russian Academy of Sciences. St. Petersburg
State University. Saint Petersburg, Russia.
http://nuclpc1.phys.spbu.ru/nucl/Abstracts/Frontiers_2005.pdf

[6] The Possible Existence of a New Particle: the Tunneling Pion. [Book of abstracts; p. 348]
'LV National Conference on Nuclear Physics' (2005, June 28-July 1). FRONTIERS IN THE
PHYSICS OF NUCLEUS. Moon, R., Vasiliev, V. Russian Academy of Sciences. St. Petersburg
State University. Saint Petersburg, Russia.
http://nuclpc1.phys.spbu.ru/nucl/Abstracts/Frontiers_2005.pdf

[7] The Possible Existence of a New Particle in Nature: the Neutral Pentaquark. [Book of abstracts; p.
349] 'LV National Conference on Nuclear Physics' (2005, June 28-July 1). FRONTIERS IN THE
PHYSICS OF NUCLEUS. Vasiliev, V. Moon, R. Russian Academy of Sciences. St. Petersburg
State University. Saint Petersburg, Russia.
http://nuclpc1.phys.spbu.ru/nucl/Abstracts/Frontiers_2005.pdf

[8] The Experiment that discovered the Photon Acceleration Effect. [Book of abstracts; p. 77]
'International Symposium on Origin of Matter and the Evolution of Galaxies' (2005, Nov 8-11).
Gridnev, K., Moon, R., Vasiliev, V. New Horizon of Nuclear Astrophysics and Cosmology.
University of Tokyo, Japan.
https://meetings.aps.org/Meeting/SES05/Content/273
https://flux.aps.org/meetings/bapsfiles/ses05_program.pdf

[9] The Conservation of Lepton Number. [Paper presentation]
'American Physical Society 72nd Annual Meeting of the Southeastern Section of the APS' (2005,
Nov 10-12). Moon, R., Calvo, F., Vasiliev, V. Gainesville, FL. USA. APS Session BC
Theoretical Physics I, BC 0008
https://meetings.aps.org/Meeting/SES05/Content/273
https://flux.aps.org/meetings/bapsfiles/ses05_program.pdf

[10] The Vortex Theory and the Photon Acceleration Effect. [Paper presentation]
'American Physical Society; March Meeting; Topics in Quantum Foundations' (2006, March 13-17). Gridnev, K., Moon, R., Vasiliev, V. Baltimore, Maryland. USA.
Abstract ID: BAPS.2006.Mar.B40.6
https://meetings.aps.org/Meeting/MAR06/Session/B40.6
http://meetings.aps.org/link/BAPS.2006.MAR.B40.6

[11] The St Petersburg State University experiment that discovered the Photon Acceleration Effect.
'American Physical Society; March Meeting' GENERAL POSTER SESSION (2006, March 13-17).Gridnev, K., Moon, R., Vasiliev, V. Baltimore, Maryland. USA.
Abstract ID: BAPS.2006.MAR.Q1.146
https://meetings.aps.org/Meeting/MAR06/Session/Q1.146
http://meetings.aps.org/link/BAPS.2006.MAR.Q1.146

[12] The Neutral Pentaquark.
'American Physical Society; March Meeting' GENERAL POSTER SESSION (2006, March 13-17).Moon, R., Calvo, F., Vasiliev, V. Baltimore, Maryland. USA.
Abstract ID: BAPS.2006.MAR.Q1.147
https://meetings.aps.org/Meeting/MAR06/Session/Q1.147
http://meetings.aps.org/link/BAPS.2006.MAR.Q1.147

[13] The Neutral Pentaquark. [Paper presentation]
'International Workshop on "Nuclear Physics with RIBF' (2006, March 13-17).
Vasiliev, V., Calvo, F., Moon, R. RIKEN Research Institution. Saitama, JAPAN.
Abstract: RIBF-Pentaquark.
https://ribf.riken.jp/RIBF2006/

[14] Nuclear Structure and the Vortex Theory. [Paper presentation]
'International Workshop on "Nuclear Physics with RIBF' (2006, March 13-17).
Moon, R., Vasiliev, V. R. RIKEN Research Institution. Saitama, JAPAN.
Abstract RIBF-Vortex
https://ribf.riken.jp/RIBF2006/

[15] Experiment that Discovered the Photon Acceleration Effect. [Paper presentation]
'International Workshop on "Nuclear Physics with RIBF' (2006, March 13-17).
Moon, R., Vasiliev, V. R. RIKEN Research Institution. Saitama, JAPAN.
Abstract Moon 1
https://ribf.riken.jp/RIBF2006/

[16] To the Photon Acceleration Effect. [Paper presentation]
'APS/AAPT/SPS Joint Spring Meeting' (2006, March 21-23).
Moon, R. San Angelo, Texas. USA. Abstract ID: BAPS.2006.TSS.POS.8
https://meetings.aps.org/Meeting/TSS06/Session/POS.8
http://meetings.aps.org/link/BAPS.2006.TSS.POS.8

[17] The Saint Petersburg State University Experiment that discovered the Photon Acceleration Effect. [Paper presentation] 'American Physical Society; Astroparticle Physics II' (2006, April 22-25).
Gridnev, K., Moon, R., Vasiliev, V. Dallas, TX. USA. Abstract ID: BAPS.2006.APR.J7.6
https://meetings.aps.org/Meeting/APR06/Session/J7.6
http://meetings.aps.org/link/BAPS.2006.APR.J7.6

[18] The Photon Acceleration Effect. [Paper presentation]
'American Physical Society; Session W9 DNP: Nuclear Theory II' (2006, April 22-25).
Gridnev, K., Moon, R., Vasiliev, V. Dallas, TX. USA. Abstract ID: BAPS.2006.APR.W9.6
https://meetings.aps.org/Meeting/APR06/Session/W9.6
http://meetings.aps.org/link/BAPS.2006.APR.W9.6

[19] The Neutral Pentaquark. [Paper presentation]
'American Physical Society; Session W9 DNP: Nuclear Theory II' (2006, April 22-25). Moon, R., Calvo, F., Vasiliev, V. Dallas, Texas. USA. Abstract ID: BAPS.2006.APR.W9.9
https://meetings.aps.org/Meeting/APR06/Session/W9.9
http://meetings.aps.org/link/BAPS.2006.APR.W9.9

[20] Controversy surrounding the Experiment conducted to prove the Vortex Theory. [Paper presentation] 'American Physical Society; 8th Annual Meeting of the Northwest Section' (2006, May 18-20). Vasiliev, V., Moon, R. University of Puget Sound. Tacoma, Washington. USA. Abstract ID: BAPS.2006.NWS.C1.9
https://meetings.aps.org/Meeting/NWS06/Content/518
https://meetings.aps.org/Meeting/NWS06/Session/C1.9

[21] The Photon Acceleration Effect. [Paper presentation]
'American Physical Society; 8th Annual Meeting of the Northwest Section' (2006, May 18-20). Moon, R., Vasiliev, V. University of Puget Sound. Tacoma, Washington. USA. Abstract ID: BAPS.2006.NWS.C1.8
https://meetings.aps.org/Meeting/NWS06/Content/518
https://meetings.aps.org/Meeting/NWS06/Session/C1.8
http://meetings.aps.org/link/BAPS.2006.NWS.C1.8

[22] Experiment that Discovered the Photon Acceleration Effect. [Paper presentation]
'International Congress on Advances in Nuclear Power Plants' ICAPP '06, (2006, June 4-8). Gridnev, K., Moon, R. Reno, Nevada. USA. American Nuclear Society.
Abstract 6006. ISBN: 978-0-89448-698-2

[23] The Neutral Pentaquark. [Paper presentation]
'International Congress on Advances in Nuclear Power Plants' ICAPP '06 (2006, June 4-8). Vasiliev, V., Calvo, F., Moon, R. Reno, Nevada. USA. American Nuclear Society.
Abstract 6045. ISBN: 978-0-89448-698-2

[24] Is Hideki Yukawa's explanation of the strong force correct?
'The International Symposium on Exotic Nuclei' Book of abstracts: Joint Institute for Nuclear Research. (2006, July 17-22). Vasiliev, V., Moon, R. Khanty Mansiysk, Siberia. Russia.
http://wwwinfo.jinr.ru/exon2006/
http://jinr.ru/

[25] The Explanation of the Pauli Exclusion Principle. [Paper presentation]
'59th Annual meeting of the American Physical Society Division of Fluid Dynamics' (2006, Nov 19-21). Moon, R., Vasiliev, V. Tampa, Florida. USA. American Physical Society;
Abstract ID: BAPS.2006.DFD.P1.17
https://meetings.aps.org/Meeting/DFD06/Content/578
https://meetings.aps.org/Meeting/DFD06/Session/P1.17
http://meetings.aps.org/link/BAPS.2006.DFD.P1.17

[26] Is Hideki Yukawa's explanation of the strong force correct? [Paper presentation]
'59th Annual meeting of the American Physical Society Division of Fluid Dynamics' (2006, Nov 19-21). Moon, R., Vasiliev, V. Tampa, Florida. USA. American Physical Society;
Abstract ID: BAPS.2006.DFD.P19
https://meetings.aps.org/Meeting/DFD06/Content/578
https://meetings.aps.org/Meeting/DFD06/Session/P1.19
http://meetings.aps.org/link/BAPS.2006.DFD.P1.19

[27] The Final Proof of the Michelson Morley Experiment; The explanation of Length Shrinkage and Time Dilation. [Book of materials] 'Scientific Research Center for Ecological Safety of the Russian Academy of Sciences: Scientific Seminar of Ecology and Space'. (2007, February 8-10). Moon, R. Saint Petersburg, Russia.
https://spcras.ru/en/srcesras/

[28] The Explanation of the Photon's Electric and Magnetic fields and its Particle and Wave Characteristics. [Paper presentation] 'Annual Meeting of the Division of Nuclear Physics Volume 52, Number 10'. (2007, Oct 10-13). Moon, R., Vasiliev, V. Newport News, Virginia. USA. American Physical Society; Abstract ID: BAPS.2007.DNP.BF.15
https://meetings.aps.org/Meeting/DNP07/Session/BF.15
http://meetings.aps.org/Meeting/DNP07
http://meetings.aps.org/link/BAPS.2007.DNP.BF.15

[29] The St. Petersburg State University experiment that discovered the Photon Acceleration Effect. 'Virtual Conference on Nanoscale Science and Technology' VC-NST. (2007, Oct 21-25). Moon, R., Vasiliev, V. University of Arkansas. 222 Physics Building. Fayetteville, AR 72701 USA.
http://www.ibiblio.org/oahost/nst/index.html

[30] The Explanation of Quantum Teleportation and Entanglement Swapping. [Paper presentation] '49th Annual Meeting of the Division of Plasma Physics, Volume 52, Number 11' (2007, Nov 12–16). Moon, R., Vasiliev, V. Orlando, Florida. American Physical Society;
Abstract ID: BAPS.2007.DPP.UP8.21
https://meetings.aps.org/Meeting/DPP07/Content/901
http://meetings.aps.org/link/BAPS.2007.DPP.UP8.21
https://meetings.aps.org/Meeting/DPP07/Session/UP8.21

[31] The Explanation of the Photon's electric and magnetic fields, and its particle and wave characteristics. [Paper presentation]
'60th Annual Meeting of the Division of Fluid Dynamics'. Volume 52, Number 12. (2007, Nov 18–20). Moon, R., Vasiliev, V. Salt Lake City, Utah. American Physical Society;
Abstract ID: BAPS.2007.DFD.JU.22
http://meetings.aps.org/Meeting/DFD07
https://meetings.aps.org/Meeting/DFD07/Session/JU.22
http://meetings.aps.org/link/BAPS.2007.DFD.JU.22

[32] The Explanation of quantum entanglement and entanglement swapping. [Poster Session] 'The 10th International Symposium on the Origin of Matter and the Evolution of the Galaxies (OMEG07) (2007, Dec 4-6) Moon, R., Vasiliev, V. Hokkaido University, Sapporo, Japan. Bibcode: 2008AIPC.1016.....S, Harvard (Astrophysics Data System) ISBN 0735405379
https://ui.adsabs.harvard.edu/abs/2008AIPC.1016.....S/abstract

Books by author {A} and work presented in other published books/booklets

1. *"The Vortex Theory of Matter"* Copyright 1990
   R. Moon. {A} Costa Mesa, California

2. *"The End of The Concept of Time"* Copyright 2000.
   R. Moon. {A} Gordon's Publications of Baton Rouge. Louisiana. ISBN 096792981-4.

3. *"The Bases of the Vortex Theory of Space"* (2002).
   R. Moon. {A} Publishing house; "ZNACK" Director Dr. I. S. Slutskin. Post Office Box 648. Moscow, 101000, Russia. p. 32. (In Russian). Journal ISSN: 2362945.

4. *"The Vortex Theory…The Beginning"* (2003). Copyright 2003.
   R. Moon. {A} (Editor's note by Prof., Dr. Victor V. Vasiliev)
   Gordon's Publications of Fort Lauderdale Fla. USA.

5. *"The Bases of the Vortex Theory"* (2003).
   Book of abstracts: Russian Academy of Sciences; ISBN 5-98340-004-5; TRN: RU0403918096768 OSTI ID: 20530263 R. p. 251. R. Moon. V. Vasiliev
   http://nuclpc1.phys.spbu.ru/nucl/Abstracts/Nucleus_2003.pdf
   http://physics.doi-vt1053.com/ISBN5-98340-004-5/Nucleus_2003.pdf

6. *"The Vortex Theory and some interactions in Nuclear Physics"* (2004).
   Book of abstracts: Russian Academy of Sciences; ISBN 5-9571-0075-7 p. 259.
   R. Moon. V. Vasiliev
   http://nuclpc1.phys.spbu.ru/nucl/Abstracts/Nucleus_2004.pdf
   http://physics.doi-vt1053.com/ISBN5-9571-0075-7/Nucleus_2004.pdf

7. The Vortex Theory Explains the Quark Theory. (2005).
   R. Moon. {A} Gordon's Publications of Fort Lauderdale, Florida. USA. p. 205.

8. Dr. Russell Moon PhD Thesis; *"The End of "Time"* Collection of Learned Works Addendum, 2012, (pp. 473-488) VVM Publishing House: ISBN 978-5-9651-0804-6 Editor in Chief: I. S. Ivlev. Saint Petersburg State University. St Petersburg, Russia.
   http://physics.doi-vt1053.com/ISBN978-5-9651-0804-6/Dr-Russell-G-Moon-PhD-thesis-The-End-of-Time.pdf
   http://physics.doi-vt1053.com/ISBN978-5-9651-0804-6/Natural_Anthropogenic_Aerosoles_4pages.pdf

9. *"The Discovery of the Fifth Force in Nature: The Anti-Gravity Force"* Collection of Learned Works (pp. 489-495) R. Moon. M. F. Calvo.
   VVM Publishing House: ISBN 978-5-9651-0804-6 p. 534. Editor in Chief: I. S. Ivlev. St Petersburg State University. St Petersburg, Russia.
   http://physics.doi-vt1053.com/ISBN978-5-9651-0804-6/The-Discovery-of-the-Fifth-Force-in-Nature:-The-Anti-gravity-Force.pdf
   http://physics.doi-vt1053.com/ISBN978-5-9651-0804-6/Natural_Anthropogenic_Aerosoles_4pages.pdf

10. *"The Discovery of the Fifth Force in Nature: The Anti-gravity Force"* Collection of Learned Works (pages 496-503) V. Vasiliev. R. Moon. M. F. Calvo. VVM Publishing House; ISBN 978-5-9651-0804-6 2013. p 534. Editor-in-Chief: I. S. Ivlev, Saint Petersburg State University. St. Petersburg. Russia.
http://physics.doi-vt1053.com/ISBN978-5-9651-0804-6/The-Discovery-of-the-Fifth-Force-in-Nature:-The-Anti-gravity-Force.pdf

Other References

1) Christopher Scarre. Smithsonian Institution; '*Smithsonian Timelines of the Ancient World'*. p 65. Published September 15th 1993. ISBN-10: 1564583058

2) H. Yukawa, "Tabibito" (The traveler) World Scientific, Singapore. (1982) pp. 190-202. ISBN-10 9971950103

3) Wolfgang Pauli, Nobel Lecture; *for the discovery of the exclusion principle.* Stockholm, Sweden, (1946).
https://www.nobelprize.org/uploads/2018/06/pauli-lecture.pdf

4) Robert Desbrandes, Daniel Van Gent, '*Intercontinental quantum liaisons between entangled electrons in ion traps of thermoluminescent crystals'*. (2006-11-09), arXiv:quant-ph/0611109.
https://arxiv.org/ftp/quant-ph/papers/0611/0611109.pdf
https://doi.org/10.48550/arXiv.quant-ph/0611109

5) D. Bouwmeester, J.-W. Pan, K. Mattle, M. Eibl, H. Weinfurter, A. Zeilinger, '*Experimental Quantum Teleportation',* Nature 390, 6660, 575-579 (1997), arXiv:1901.11004.
https://doi.org/10.48550/arXiv.1901.11004

6) Einstein, A. and Podolsky, B. and Rosen, N., '*Can Quantum-Mechanical Description of Physical Reality Be Considered Complete?'* Phys. Rev. 47, 777, (1935).
https://link.aps.org/doi/10.1103/PhysRev.47.777
https://journals.aps.org/pr/abstract/10.1103/PhysRev.47.777

7) Rucker, Rudy. '*The Fourth Dimension: Toward a Geometry of Higher Reality'.* (1984) ISBN-10: 0486779785

8) Abbott, Edwin, A. *FLATLAND*: '*A Romance of Many Dimensions'.* New York, Dover Publications. (1953) ISBN-13: 9798630248015

9) Besancon, Robert M. '*The Encyclopedia of Physics'.* pp 568-570; and 949-950. Van Nostrand Reinhold Co. (1974) ISBN-10: 1124004475

10) Condon, E. U. '*The Handbook of Physics Second Edition'.* McGraw Hill Book Co. (1967) ISBN-10: 0070124035

11) Eisberg, Robert. '*Fundamentals of Modern Physics'.* Pages 9-15. John Wiley and Sons. (1961) ISBN-10: 047123463X

12) Dorothy Michelson Livingston. '*The Master of Light'.* A biography of Albert A. Michelson. University of Chicago Press. (1973) ISBN-10: 0684134438

Russian Scientific Journals:

1. http://www.new-philosophy.narod.ru/RGM-VVV-RU.htm (in Russian)

2. http://www.new-philosophy.narod.ru/MV-2003.htm (in English)

3. http://www.new-idea.narod.ru/ivte.htm (in English)

4. http://www.new-idea.narod.ru/ivtr.htm (in Russian)

5. http://www.new-philosophy.narod.ru/mm.htm (in Russian)

The End?